STYLISH STORES II

时尚专卖店 II

VOL. 2

深圳市艺力文化发展有限公司 编

·广州·

图书在版编目（CIP）数据

时尚专卖店 II = Stylish stores II ：英文 / 深圳市艺力文化发展有限公司编. — 广州 ：华南理工大学出版社，2014. 3
ISBN 978-7-5623-4176-5

Ⅰ. ①时… Ⅱ. ①深… Ⅲ. ①商店－室内装饰设计－作品集－世界－英文 Ⅳ. ① TU247.9

中国版本图书馆 CIP 数据核字（2014）第 038111 号

时尚专卖店 Ⅱ　Stylish stores Ⅱ
深圳市艺力文化发展有限公司 编

出 版 人：**韩中伟**
出版发行：华南理工大学出版社
（广州五山华南理工大学 17 号楼，邮编 510640）
http://www.scutpress.com.cn　E-mail: scutc13@scut.edu.cn
营销部电话：020-87113487　87111048（传真）
策划编辑：赖淑华
责任编辑：王岩
印 刷 者：深圳市新视线印务有限公司
开　　本：635mm×1020mm 1/8　**印张**：55
成品尺寸：245mm×300mm
版　　次：2014 年 3 月第 1 版　2014 年 3 月第 1 次印刷
定　　价：598.00 元（共 2 册）

PREFACE

Stores are the portal to a lifestyle we wish to appropriate. They are the opportunity to exact gratification from our demanding lives through the ultimate of capitalist rites: the purchase.

Beyond the private sphere and the marginal value of the purchase, there is the social narrative of shopping, often shared with people who are, for different reasons, significant to that shopping experience. Brands are aware of the increased value of their products once it is offered within the larger message of lifestyle. Shoppers are persuaded to make the purchase by the impulse/need of being part of a recognizable group.

These are the premises to the design process of a store. The architect and the designer are tasked to define the spatial realm of a sequence of messages, and they do so with different approaches, attitudes, aptitudes… though, all with the intent of complementing and/or enriching the experience of purchasing.

The continual proliferation of brands, also due to the lowering of manufacturing costs, has compelled brand owners to offer evermore unique design and experiential values… which are used as a source of distinction and identity. Architects and designers have seized this opportunity to create spaces which are formally, technically, and technologically at the forefront of architectural innovation and research.

The projects in the following pages portend an ever-evolving scenario of exciting developments. This book is an indispensable compendium of outstanding retail venues from across the globe. It showcases remarkable speculative reach and technological advancements which deliver extraordinary formal and experiential values.

Antonio Di Oronzo, principal

bluarch architecture + interiors + lighting

Professor of architecture [City College of New York – Spitzer School of Architecture]

CONTENTS

BEAUTY STORES

BOOK & STATIONERY STORES

OPTICAL STORES

PHARMACIES

FOOD & DRINK STORES

HOUSEHOLD STORES

JEWELRY STORES

ELECTRIC STORES

OTHERS

BEAUTY STORES

"Crème de la Crème" Haute Parfumerie

Design Agency:
INBLUM architects

Design Team:
Laura Malcaitė, Dmitrij Kudin

Client:
Crème de la Crème

Location:
Klaipėda, Lithuania

Photography:
Darius Petrulaitis

AMOUAGE
Crème de la Crème
Crème de la Crème
Crème de la Crème

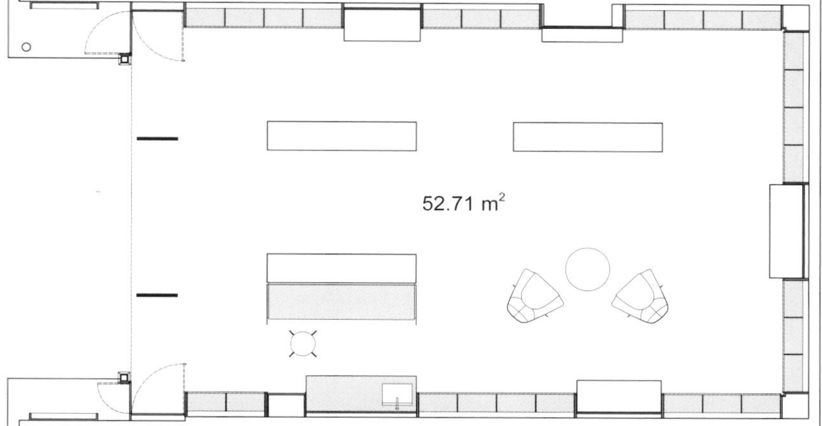

Just imagine a library that doesn't essentially consist of books, but is full of delicate personal objects, dear reminiscences, odours, yet unveils links and feelings of future experiences.

A vision that once emerged from dreams as an archetype became an invisible frame for the whole "library" to be mounted on.

Designers have been looking for a close balance between ephemerality and monumentality, industry and luxury, aesthetic constants and present-day advance. Should an interior be imagined as a scent, they would want it to be intellectual, refined. It is the search for intellectual refinement that encouraged unusual precision and greater focus on detail, careful selection of materials and their combinations. Eventually, one glance may not be sufficient to observe the transition of the structural solutions into delicate interior details, while the archetypical image of the library has acquired forms that spread harmony and intellect.

The linear structure has integrated the interior into a visually rhythmic completeness, where it is easy to find oneself and conceive the exposition sequence. Each brand occupies a separate shelving segment marked with a corresponding book above. The dominating trademarks are vividly presented, their names and logos embossed on impressive veneered panels. The carefully selected wood texture pattern and the glossy surface produce a row of warm hues, which, together with black-painted aluminium shelves add to an impression of depth and luxury.

This impression is also enforced by the lighting systems. The central light beam is directed to the shelves housing rows of sparkling little bottles resembling actors on stage. Meanwhile the inner space is sunk into the pleasant twilight with only a few objects shining, the shelving islets, and, an essential element of the interior, a sculptural group of spongy bronze chandeliers hanging over the cash desk.

The hanging golden swarm reflects something on perfection, seduction... As if striving to arouse instincts, yield to the pleasures of smell... in the library, among "books", whose texts are odours.

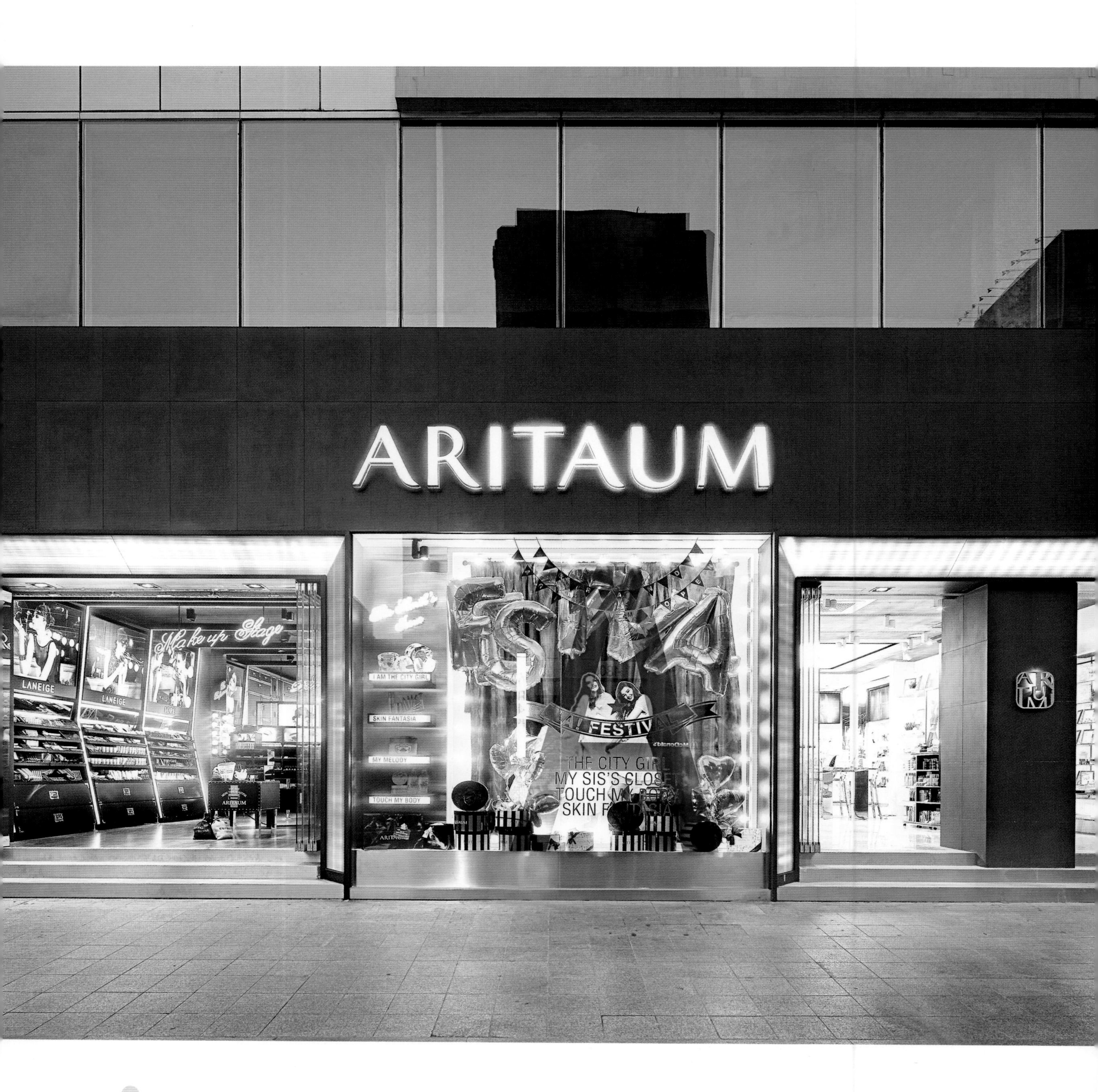

Aritaum Flagship Store

Design Agency:
URBANTAINER Co., Ltd.

Design Team:
Younjin Jeong, Daesoon Hwang, Semi Kim

Client:
AMOREPACIFIC Coporation

Area:
165 m^2

Photography:
Sun Namgoong

LANEIGE
LANEIGE
IOPE
Mamonde

LANEIGE
LANEIGE
IOPE

ARITAUM flagship store at Gang-Nam station is interesting place where existing image of Asian beauty is maintained well as a symbol of trend and elegance in the design of ARITAUM.

The interior design which reflects Traditional Korean house is familiar, but somewhat fresh to URBANTAINER.

The spaces are largely classified as 4 zones: MAKE-UP STAGE ZONE where traditional column divisions used, BROW & NAIL BAR ZONE where URBANTAINER's comfortable resting place of main floored room was represented, ARITAUM SQUARE ZONE where concept of open space like as leaving the door open to share space, and BEAUTY SOLUTION ZONE where women have friendly talked.

This project has completed by collaboration of elegant traditional beauty, trendy concept of Asian beauty, and the designer's challenging idea of "Classic is chic". From stylish and refined sense onto the monotonous traditional construction, newly innovative brand image of ARITAUM can be achieved.

Due to the characteristic of cosmetic shop, the space is comprised of black and white to emphasize colorful make-up products and skincare products that help to achieve clean skin. Also, various concept of each zone reflect the taste of modern young women.

The furniture with a concept of make-up box in the place can have various roles such as becoming oversized table or drawer.

This transformation through one object factor plays roles as maintaining the uniformity in the space and inspiring humor.

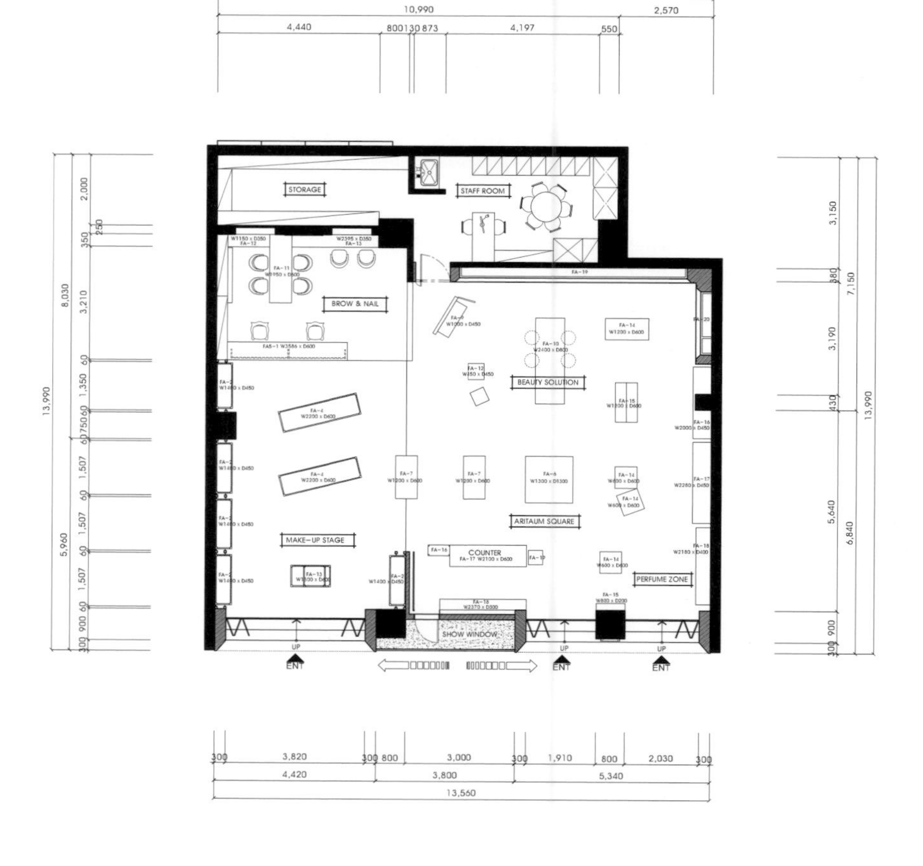

FLAGSHIP STORE
GANGNAM
ARITAUM

FLAGSHIP STORE
GANGNAM

Club Clio Professional

Design Agency:
URBANTAINER Co., Ltd.

Design Team:
Younjin Jeong, Yoon Chang

Client:
CLIO CO., LTD.

Location:
Joong Gu Myeong-Dong, Seoul, Korea

Area:
72 m^2

Photography:
Sun NamGoong

Korean cosmetics brand "Clio" opened its first flagship store "Club Clio" in Myeong-Dong. Club was chosen as the store's identity concept, large speakers and the media wall visualize a loud and colorful atmosphere while the electronic music further complements the club experience.

Upon entering "Club Clio", guests are greeted by a bar-like furniture setup that doubles as product displays. Further into the store are the front-facing LED panels and cashier register that resembles a DJ booth. Strategically installed speakers serve as iconic elements that amplify the space's club experience.

Mono-tone finish used throughout the store emphasizes Clio's colorful product line. From the cement finished walls and matte finished furniture, the Club Clio's spatial elements blend into the background while successfully shifting the focus to the bright colors of Clio.

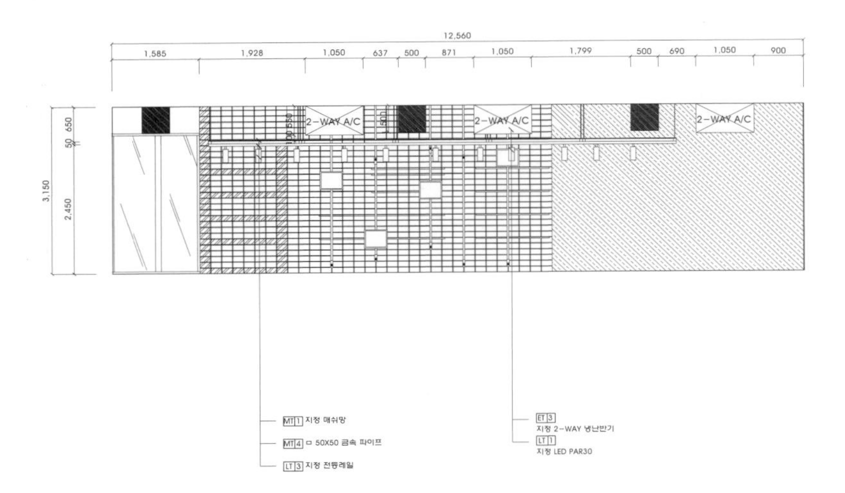

Section

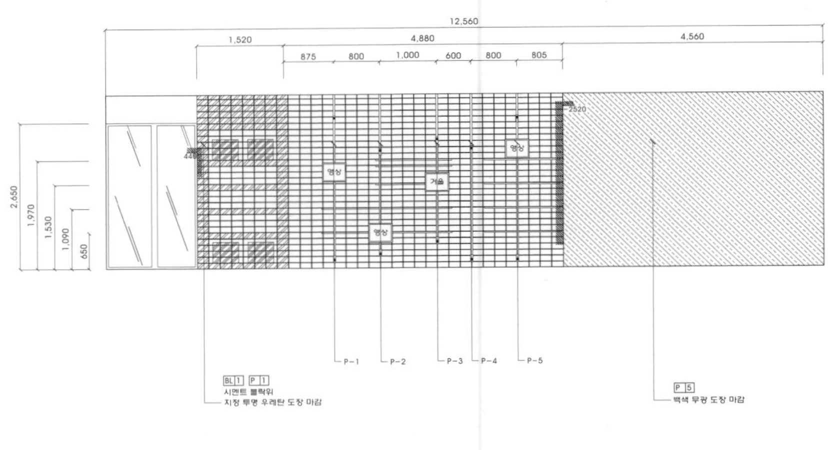

Elevation-01

TOOL
EYE

ART
TOOL
EYE

BOOK & STATIONERY STORES

BiC Wholesale Store in Shanghai

Design Agency:
Beijing Matsubara and Architects

Design Team:
Hironori MATSUBARA, Satoshi YAMADA

Location:
Shanghai, China

Area:
40 m^2

Photography:
Nacasa and Partners (Eiichi Kano)

BiC
BIC 比克铅笔
给你明星般的书写体验!
子画得开心!

It's the interior design for the first wholesales store in China of French stationery company, "BiC".

The site is located in the Chinese typical wholesale market building. The space is crowded with stationery wholesale stores.

There are a lots of similar size stores, thereby the eye-catching façade and the storage space for tons of wholesale stocks are the key of the design for this project.

Designers put the shelves on the both side walls for the enough storage space, and install angled orange walls between each shelf for avoiding the view from the entrance. Radial arrangement of shelves emphasizes the depth of the space, and makes customers focus on the orange walls that exhibit BiC products and products on them. The wall in the end was covered with mirror and reducing the oppressive feeling of the confined store space.

Once coming into the store, the shelves are not seen from the entrance since they are behind of the tilted orange walls. Their products, mainly pens, are displayed on the orange walls with different way. And in the centre of this store, there is a cloud-shaped counter which has different levels. This counter exhibits products and has cash resister also. The same shape hanged ceiling is illuminated with rainbow colored acrylic tubes.

In the shop façade, the sign of their logo "BiC" and the rainbow colored lightings are installed as the eye-catching factor.

For avoiding the smallness of the store space, the visual effects of the tilted walls and the mirrors emphasize the depth of the space.

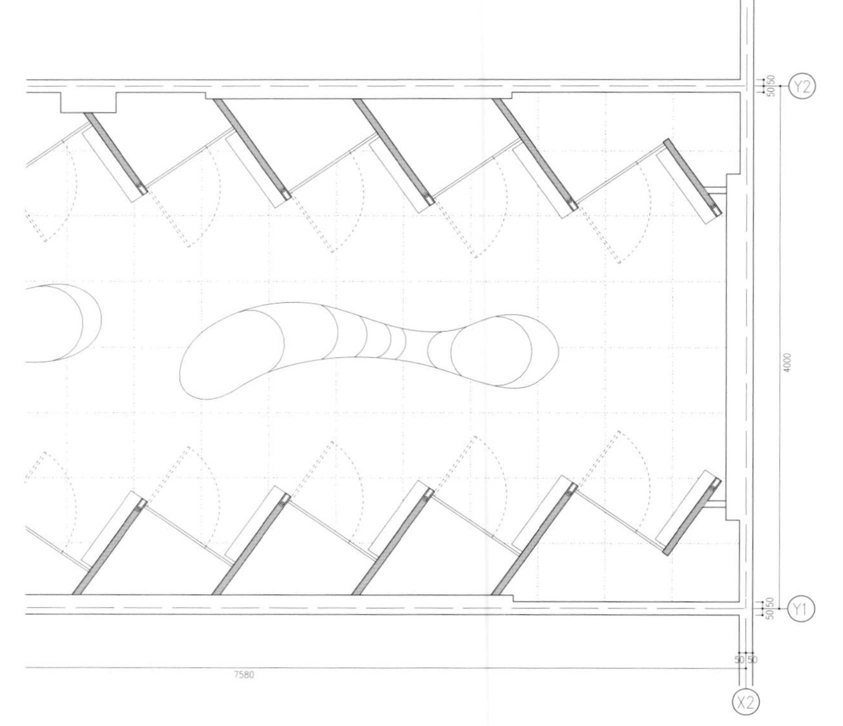

BiC

BiC

BiC 比克创意
BiC
for Her
La Vie en Rose
A NEW WAVE
比克超值系列

The Story Unfolds

Design Agency:
Thelaunchroom

Designer:
Leong Huang Zi

Client:
TSU Marketplace

Location:
Kuala Lumpur, Malaysia

THE BOOK CLUB

THE STORY UNFOLDS
THE BOOK CLUB
ART / DESIGN
BUSINESS
ARTS

The Story Unfolds (TSU) is a retail concept that allows creative designers alike to publish their own books and sell them right at the store. With a wide selection of both self-published titles as well as books by established authors.

TSU gives a sense of prestige to independent authors in how they sell and distribute their creative books. All self-published titles will be given the same treatment as their established counterparts in how they are displayed, packaged, marketed and sold at the store.

The concept of an unfolding origami is used as an analogy to tell the story behind the concept. All of us have a story to tell. It could be a story about out drawings or a story about our photos, or an actual story for a novel. TSU helps unfold them.

ARTS|ARTS
ART/DESIGN
CITYSCAPES
RETAIL ART
RETAIL ART

ART/DESIGN
Wallpaper*
GRAPHICS
RETAIL ART
FRAME
Wallpaper*
Wallpaper*
ARTS
INTERIORS
WOOD BLOCK ART

OPTICAL STORES

Magic Store

Design Agency:
TORAFU ARCHITECTS Inc.

Location:
Aoyama Tokyo, Japan

Area:
32.6 m^2

Photography:
Daici Ano

Magic
Store

Magic

TORAFU ARCHITECTS designed the interior of a concept store that opened in Omotesando, Tokyo featuring Menicon's new line of contact lenses; Magic. Using the comprehensive contact lens manufacturer's unique Flat Packaging Technology — said to be the world's thinnest (1mm) — Magic represents a major departure from conventional contact lens packaging.

TORAFU ARCHITECTS proposed a ribbed display wall where the thin and compact packages can be shelved on any part of it. The wall, which wraps the interior of the store smoothly, has extrusions and depressions matching the size of these packages — like a white canvass on which to display products freely. By filling the whole display wall with Magic casings, the store is made to look like one big showcase room that faces the street. Moreover, the wall's irregular surface create a variety of expressions of light and shadows and the plants set into the ceiling give the impression the store has been turned upside-down.

TORAFU ARCHITECTS aimed to create a showcase-like space that gives off a feeling of novelty every time one looks at the ever-expanding color variation of packages on display.

Magic
Magic

OPTICON Hamburg

Design Agency
GRAFT Gesellschaft von Architekten mbH

Design Team:
Sven Fuchs, Gunhild Niggemeier, Julian Busch, Christoph Jantos, Alejandra Lillo

Client:
Hamburg Eyewear on GmbH

Location:
Hamburg, Germany

Area:
140 m^2

Photography:
Christian Barz and Felsch Lighting Design GmbH

A feeling of wellbeing, trust, spaciousness, intimacy and flexibility are the key factors for the shop design. Rather than separating interior and exterior by a special shop window, the store stages a holistic space for display and interaction which can be read from the street when passing by.

Reconstruction in a historical building in Hamburg initiated this somehow atypical, rather adventurous and very individual store development. The large number of products on display — more than 2,100 pairs of glasses is to be presented in several changing exhibitions — have become an integral part of the shop design. The merchandise is always visible, freely accessible but never obtrusive. Presentational areas and space boundaries intersect, offset each other, a wall becomes a shelf, or a workshop, cupboard and even show case.

As an open space, vibrating within its boundaries, it remains free and is dominated by two large pieces of furniture. The sculptural wall continuum embraces customers and guests and pervades the entire ground plan. Waved areas give a dynamic to this free and open space which, with the help of mirrors, seems to continue endlessly.

Curved wall sections widen and constrict the gaps and create versatile event and shop zones, accentuating the transitions between them.

In a second step, the large items of furniture orchestrate the interior of this casket whose various zones appear to be naturally evolved. Endorsed by strong colours and velvet coverings, these objects dominate the spatial zones of this space continuum.

As final element in the design of the store, the light concept comes into play. Dynamic light behind the glasses on display accentuates the curving architecture and guides the attention, while generous illumination from the front creates the appropriate scenery for the merchandise.

TOM FORD
PRADA
Cartier
GUCC
OLIVER PEOPLES
PRADA

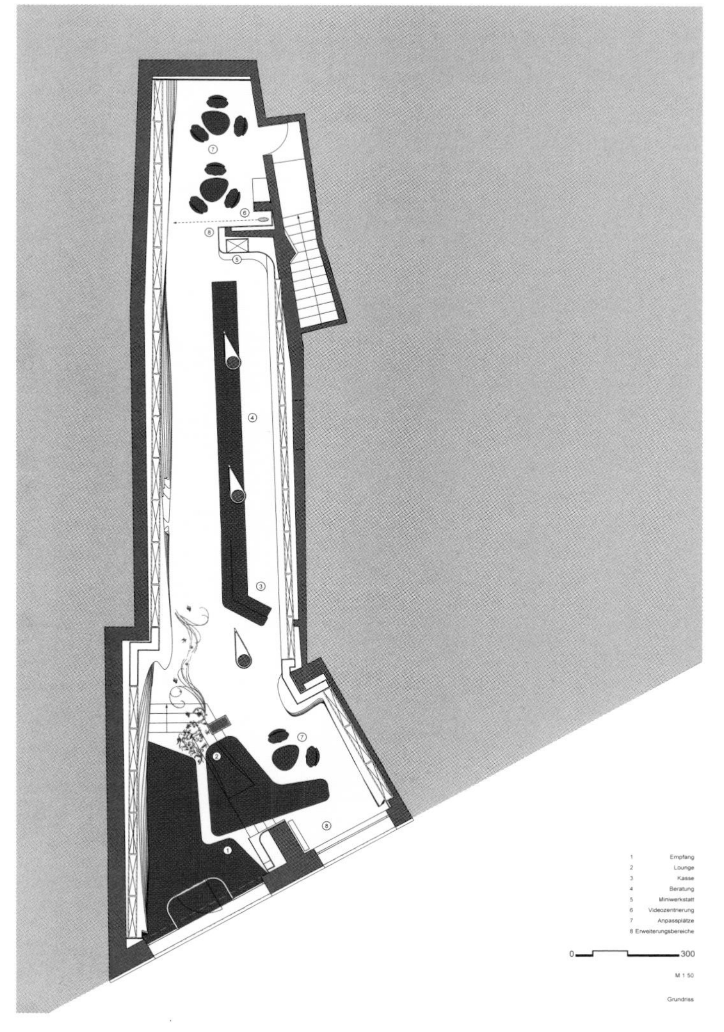

1 Empfang
2 Lounge
3 Kasse
4 Beratung
5 Miniwerkstatt
6 Videozentrierung
7 Anpassplätze
8 Erweiterungsbereiche
0
300
M 1:50
Grundriss

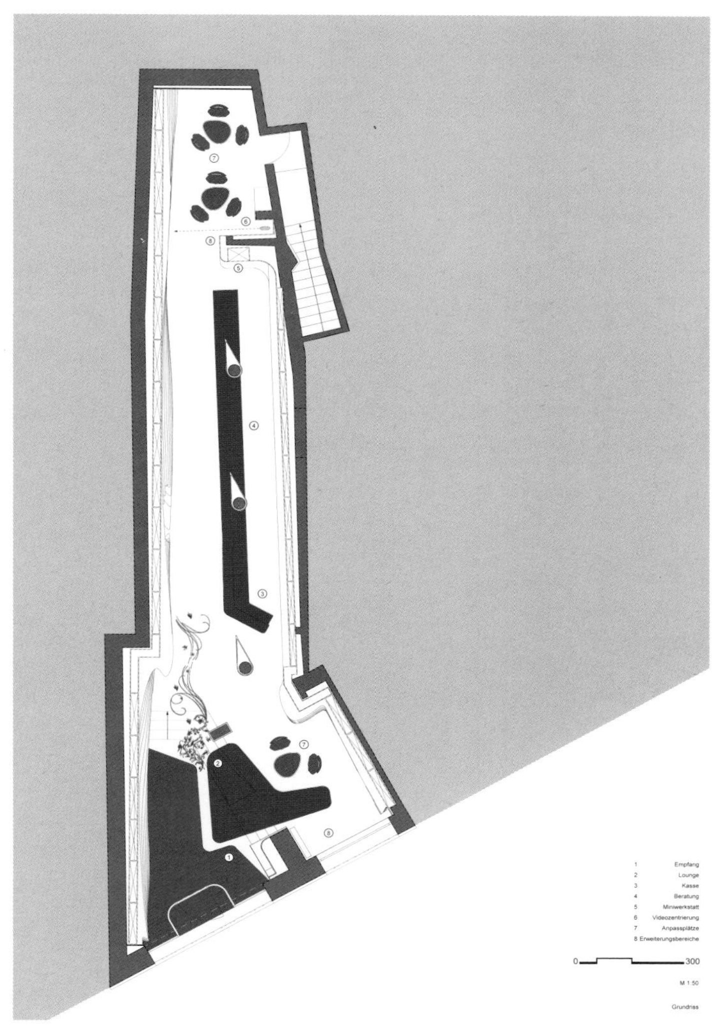

1 Empfang
2 Lounge
3 Kasse
4 Beratung
5 Miniwerkstatt
6 Videozentrierung
7 Anpassplätze
8 Erweiterungsbereiche
0
300
M 1:50
Grundriss

PHARMACIES

Farmacia Bolli 1833

Designer:
Alessia Silvestrelli

Client:
Dott. Giuseppe Cenci

Location:
Pontefelcino, Perugia, Italy

Area:
550 m^2

Photography:
Ljudmilla Socci

The pharmacy covers approximately 400 m^2 in a newly constructed building that, despite the ample size and the commercial vocation of the sites, has an internal height between floors of only 3 meters. There was a need to include, in addition to the normal technological arrangements, also the passage of the transport belt from the automatic warehouse allowing the arrival of the medicines directly to the sales counter.

Conceptually, the pharmacy, at the wish of the contractor alongside whom the complex functional layout was created, is a circular mechanism that rotates around the customer/patient. Normally, those entering are forced to scan the shelves to identify the sector of interest: cosmetic, natural, or nutrition. Here, without additional movement to the ethical counter, customers can receive the prescribed medicine thanks to the automation mechanism.

The heart of the pharmacy holds the Single Reservation Centre and an advisory area. The centrality of the area intended for services is emphasized by a series of concentric circles in the space design. This makes the lights, furnishings, and false ceiling work together to form a sort of virtual gear mechanism, ideally recalling the gears of a clock. This symbolises the time that is dedicated, with the introduction of automation into the pharmacy, to listening to and possibly advising the customer/patient, with no more looking on old shelves for boxes of prescribed medicine.

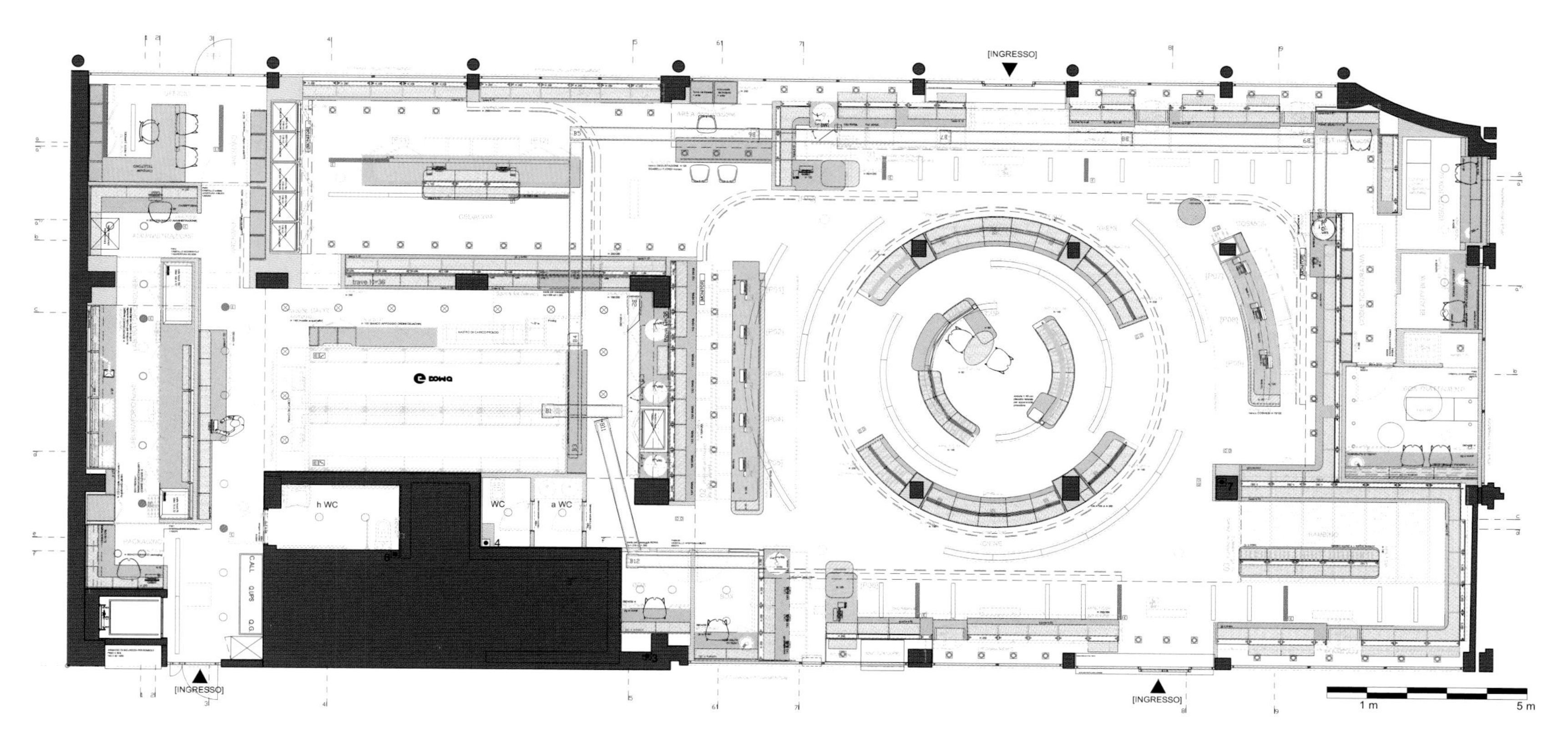
[INGRESSO]
h WC
WC
a WC
CALL
GIPS
G.G.
[INGRESSO]
[INGRESSO]
1 m
5 m

Farmacia Dorica

Designer:
Alessia Silvestrelli

Client:
Dott.ssa Simonetta Settimi

Location:
Ancona, Italy

Area:
200 m^2

Photography:
Guido Calamosca

COSMESI

The pharmacy covers approximately 180 m^2 in a newly constructed site characterised by an area overlooking an imposing entrance approximately 7 meters high, illuminated by the large windows facing the future "piazza" arrangement of the current parking area.

The idea of juxtaposition of volumes, intrinsic in the space, becomes the central theme of the project. This weaves with the constant rhythms of the vertical bands of the furnishings that modulate the perimeter, broken up by elements that escape beyond the walls to slide along the horizontal planes, such as the sign and the cosmetic counters.

The dominant colours are the tones of the Carrara marble and the glass, white, grey and crystal green that show the interpenetration of the structures and create a neutral background for displaying the products.

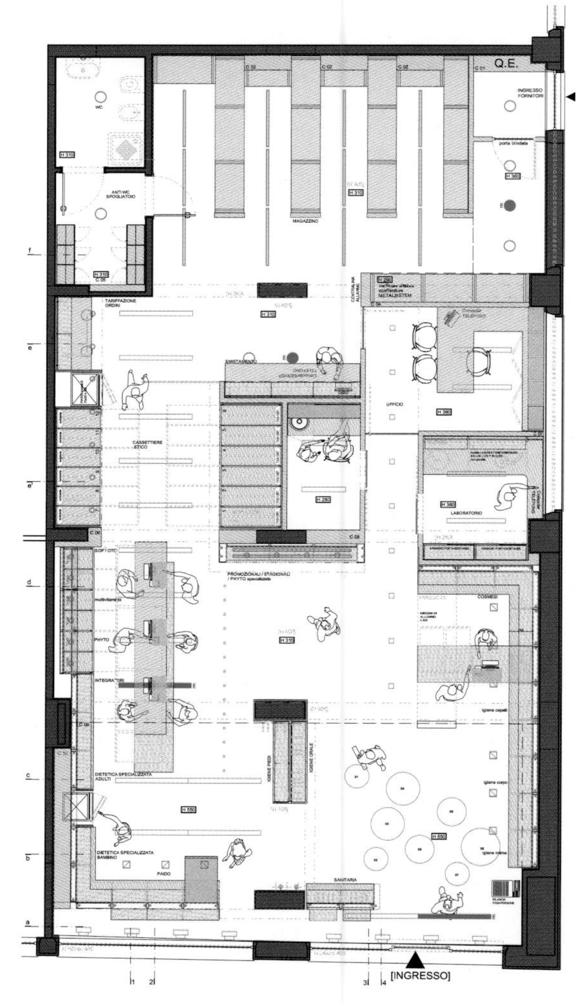

CAGNONI
03

Farmacia Giacinti

Designer:
Alessia Silvestrelli

Client:
Dott.Nello Giacinti

Location:
Grottammare,
Ascoli Piceno, Italy

Area:
250 m^2

Photography:
Enrico Maria Lattanzi

PREPARAZIONI
GALENICHE
PRESCRIZIONI
COSMESI / IGIENE

BAMBINO
ALIMENTAZIONE

PREPARAZIONI
GALENICHE
CRIZIONI
NATURALE
COSMESI / IGIENE

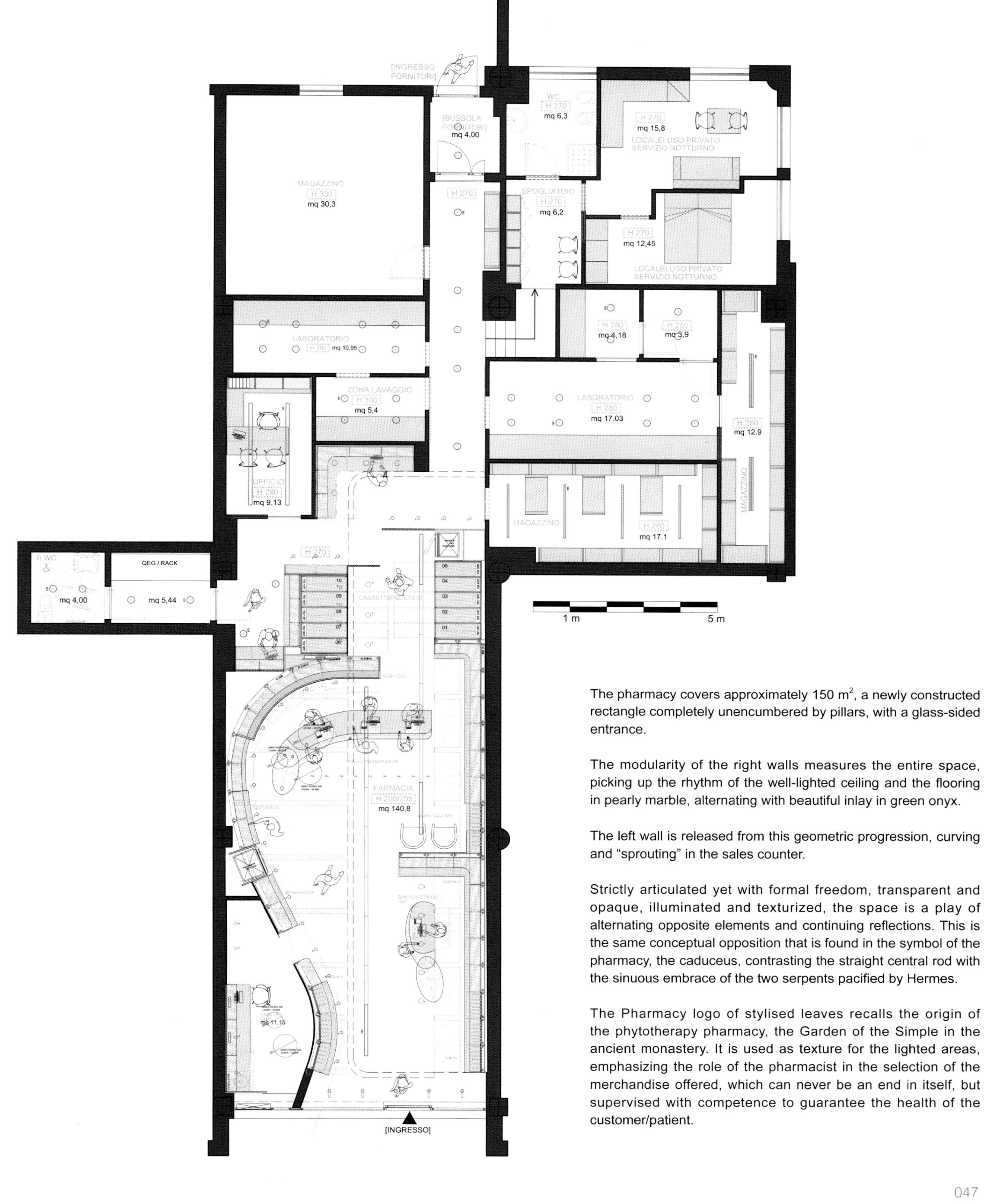

The pharmacy covers approximately 150 m^2, a newly constructed rectangle completely unencumbered by pillars, with a glass-sided entrance.

The modularity of the right walls measures the entire space, picking up the rhythm of the well-lighted ceiling and the flooring in pearly marble, alternating with beautiful inlay in green onyx.

The left wall is released from this geometric progression, curving and “sprouting” in the sales counter.

Strictly articulated yet with formal freedom, transparent and opaque, illuminated and texturized, the space is a play of alternating opposite elements and continuing reflections. This is the same conceptual opposition that is found in the symbol of the pharmacy, the caduceus, contrasting the straight central rod with the sinuous embrace of the two serpents pacified by Hermes.

The Pharmacy logo of stylised leaves recalls the origin of the phytotherapy pharmacy, the Garden of the Simple in the ancient monastery. It is used as texture for the lighted areas, emphasizing the role of the pharmacist in the selection of the merchandise offered, which can never be an end in itself, but supervised with competence to guarantee the health of the customer/patient.

Farmacia Perissinotti

Designer:
Alessia Silvestrelli

Client:
Dott.ssa Anna Perissinotti

Location:
Cordenons, Pordenone, Italy

Area:
250 m^2

Photography:
Marco Alberi Auber

TRATTAMENTI VISO / CORPO
IGIENE CAPELLI
IGIENE ORALE
IGIENE MANI / PIEDI

FARMACIA
FARMACI DA BANCO
CONSIGLIO
PRESCRIZIONI

FITOTERAPIA
IGIENE CAPELLI
IGIENE ORALE
IGIENE MANI / PIEDI

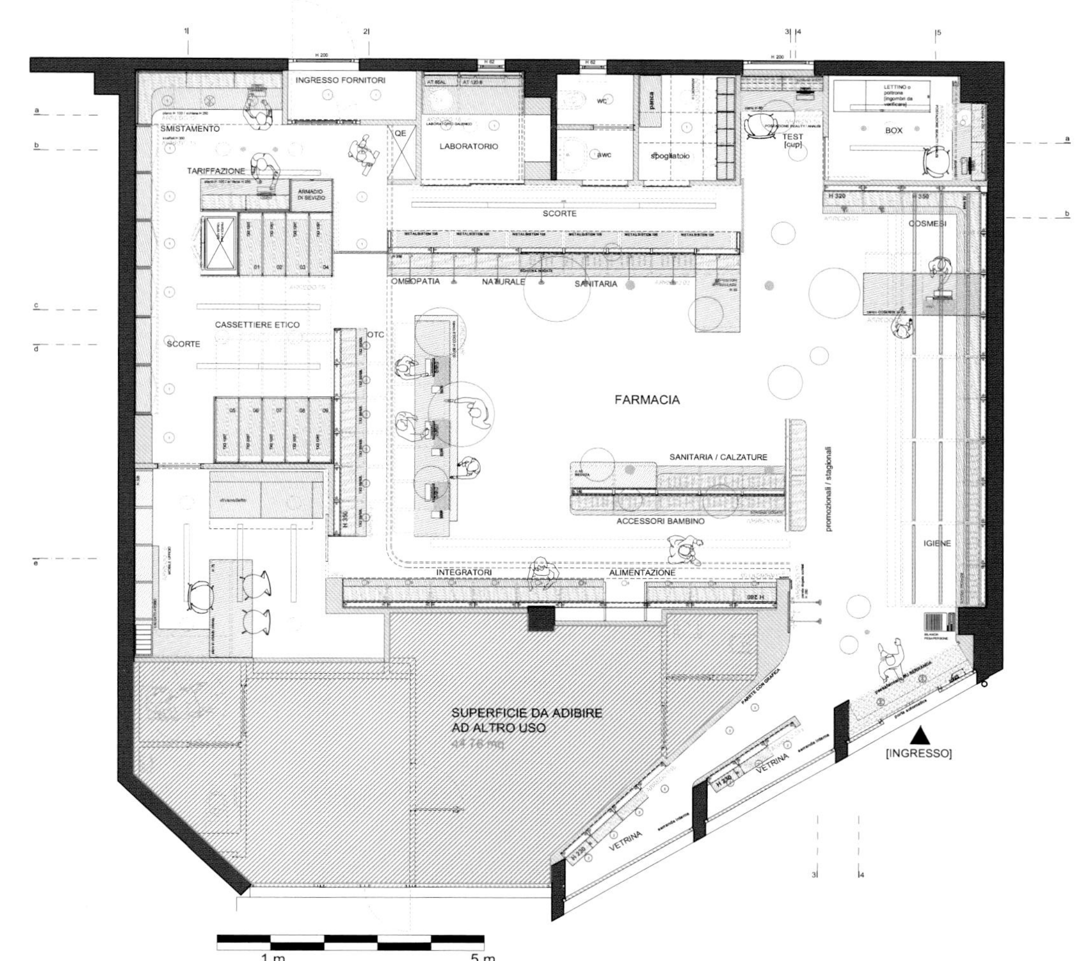

The pharmacy covers approximately 200 m^2, in part destined for medical outpatients, in a site adjacent to a shopping centre first occupied by a clothing store.

Dr. Perissinotti wanted the pharmacy to express her personality, and for it to be a comfortable and pleasant work area.

It was decided to work with neutral tones, the pearly white marble wrapping most of the surface, in contrast with a powder blue that emphasizes some elements and the intersection of the geometry, declined in different materials such as retro lacquered satiny crystal and cellular polycarbonate doors.

The entrance walls, dedicated to cosmetics, blend into the well-lighted ceiling. The strict geometry of the furnishings is softened by the casual distribution of the lights in the central plate of the ceiling, large luminous portholes that perforate the surface, released into the air like bubbles of vital oxygen.

The pharmacy is a place dedicated to the wellness of the person, ethical and therefore "aesthetic".

Patient care, the preparations made in the Galenic laboratories inside the pharmacy, is reflected in the attention to detail of the space, becoming a guarantee of reliability for the users.

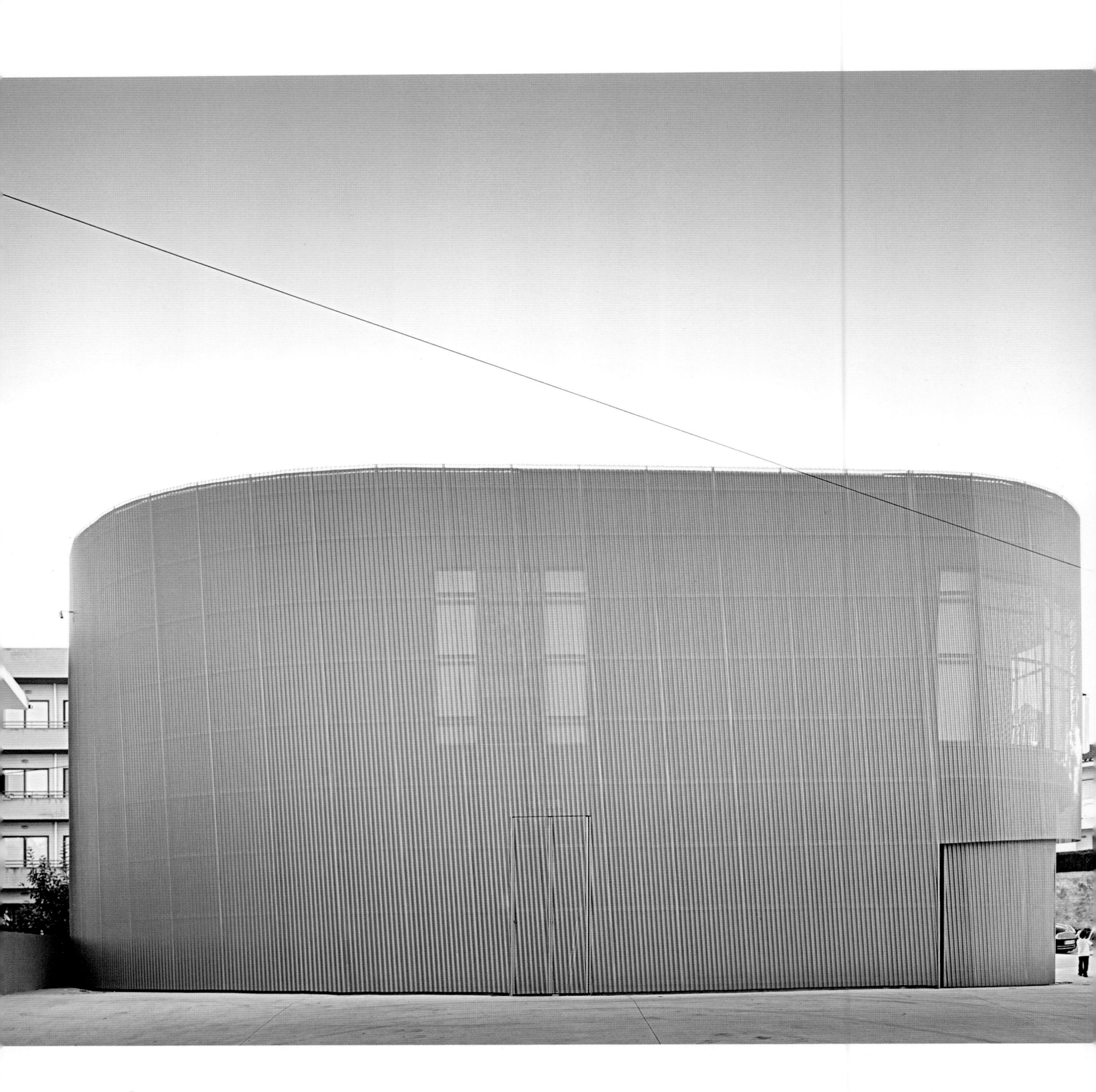

Farmácia Lordelo

Design Agency:
José Carlos Cruz - Arquitecto

Location:
Via Real, Portugal

Area:
522 m^2

Photography:
FG+SG – Fotografia de rquitectura

The pharmacy is located in Vila Real, in the centre north of Portugal and is part of a peripheral zone of the city where the environment does not have a consolidated and uniform image. In the absence of external references, it was chosen to create a building with an abstract and neutral character, reinforced by the absence of openings. With oval shape footprint, the two floors are fully aluminium coated corrugated and perforated. The only direct opening to the outside is the main entrance that gives access to the sales area. By changing the interior light and the symbol of pharmacy, the building gains dynamic, allowing the image variation from day to night. The store not only sells medicines but also has its own laboratory for compounding pharmacy.

VITAMINAS
EMAGRECIMENTO
02

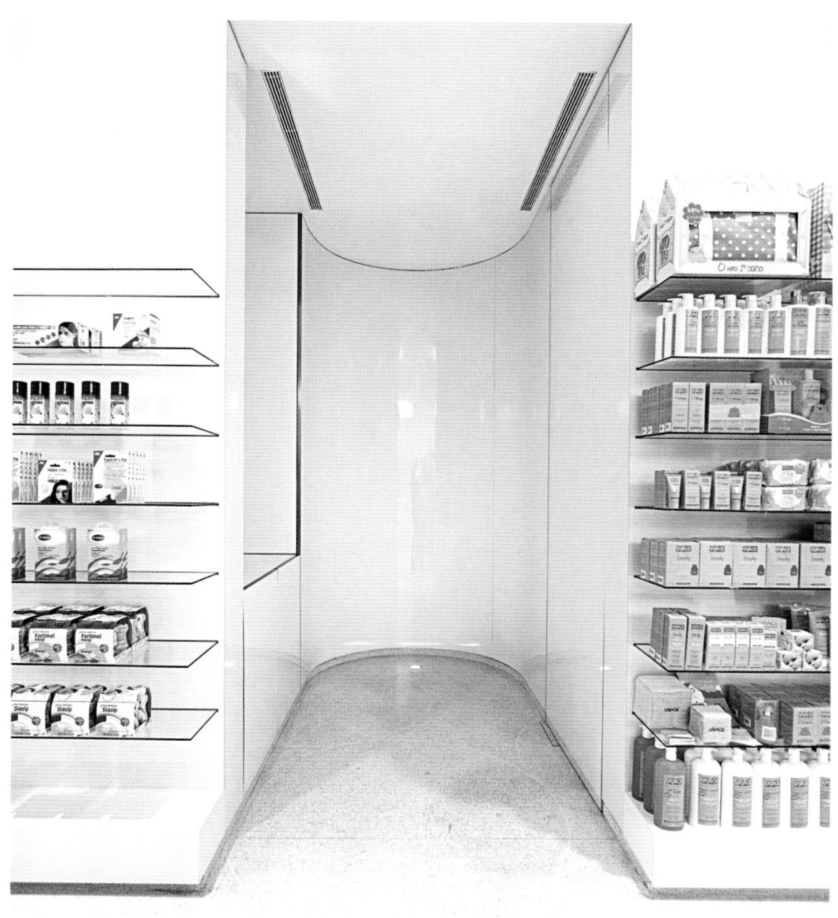

CAPILARES
HOMEM
DERMOCOSMÉTICA
OFERTA

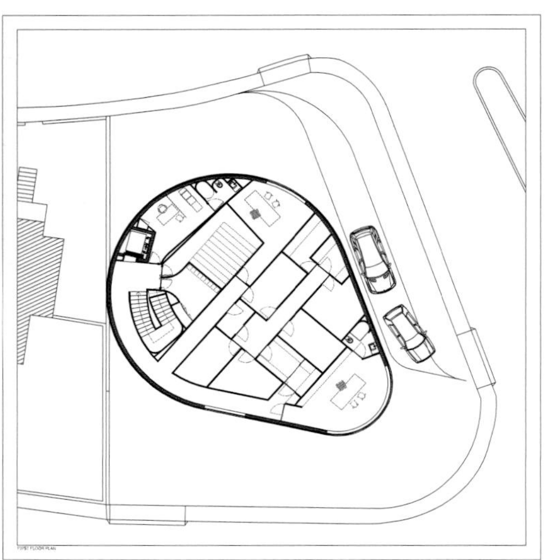
FIRST FLOOR PLAN

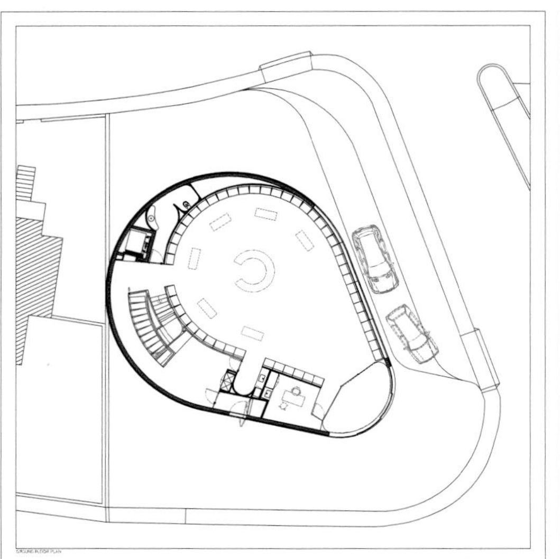
GROUND FLOOR PLAN

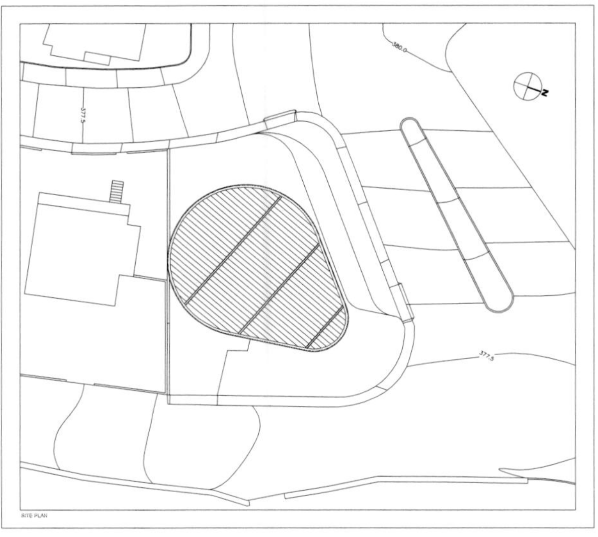
SITE PLAN

Placebo Pharmacy

Design Agency:
KLAB architects

Design Principle:
Konstantinos Labrinopoulos

Design Team:
Xara Marantidou, Enrique Ramirez, Mark Chapman

Location:
Vouliagmeni Av., Glyfada, Athens, Greece

Area:
600 m^2

Photography:
Panos Kokkinias

The design process for this large (600 m^2) supralocal pharmacy forced KLAB architects to shift their viewpoint and come up with a virtual building — a placebo pharmacy. The octagonal shape of the existing structure was re-formed into a cylinder in order to create a spiral which seeks to converse with the rapid motion on Vouliagmenis Avenue, the urban artery on which the building stands. The panels of the facade are perforated using Braille, which both alludes to the system's use on pharmaceutical packaging and boosts visibility by allowing the light to find its way into the interior. The new facade also protects the interior while acting as a lure for passers-by. Inside, the product display mirrors the circular frontage, while a ramp up to the upper level extends the dynamism of the exterior spiral into the interior space.

The pharmacy is arranged over two floors, the ground floor being the primary shop space with the upper mezzanine floor consisting of ancillary office space used as a temporary surgery for visiting health professionals.

The pharmacy is arranged in plan in a radial pattern with the main cashiers desk acting as the focal point. The product displays fan out from this focal point giving the cashier the ability to view the whole pharmacy from this central area. The drug dispensary, preparation areas and toilets are also arranged off this radial pattern. This pattern gives a natural flow to the space and allows light deep into the center of the plan at all times throughout the day.

|ΣΥΜΠΛΗΡΩΜΑΤΑ ΔΙΑΤΡΟΦΗΣ|FOOD SUPPLEMENTS|

FOOT CARE

HERMOSA Pharmacy

Design Agency:
MARKETING-JAZZ

Client:
Catalina Hermoso

Location:
Mancha Real, Jaén

Photography:
Ikuo Maruyama

200 m^2 of pharmacy, beautiful and spectacular, these are the two most repeated words by everyone who has seen it. Located in Mancha Real, a small village in the mountainous province of Jaén.

It is named after its owner, Catalina Hermoso, a person with a dream: to have the best possible pharmacy for her clients, "Everyone likes a bit of luxury" she said.

The facade would be very difficult to create in a large city or shopping centre. A 6 meters high x 6 meters wide cross and a show-stopping display of 64 backlit acrylic "albarellos", each one with a different message. Wow! Are we in Jaén or New York?

The entrance feels like you are stepping into beauty salon, two African Coralwood coated pillars with mirrors, one with a water fountain, offer a fresh and delightful welcome. A collection of circular shelves adorn all the walls in the main room. The shadows are part of the image, the product is presented to entice not to overwhelm, to tell a delectable story: your beauty is waiting to be discovered, "Try me".

The design of the room for body-care, oral hygiene and dietary products is centered on a wall with niches where the light and the colour of the wood are part of Farmacia Hermosa's unique identity. Acrylic "albarellos" and "tester" spaces form paart of the product presentation, bringing the product closer to the client.

Medicine and children's products are sold in a room christen "The garden of health". Wow! The patio of this village house has been converted into a beautiful room with an African Coralwood chequered roof, where the natural light harmonises with the pharmacy lighting.

The cash registers are albarellos and the products are presented in "flower beds". There are Provençal style waiting benches, a cabin for pharmaceutical services and a stall for the promotion of children's products. An external garden for growing medicinal plants, a ball-pit for the younger children to enjoy and an array of photographs of happy children encourage you to browse through the entire pharmacy.

"Mummy, I want to get sick so I can come back to this pharmacy", can often be heard.

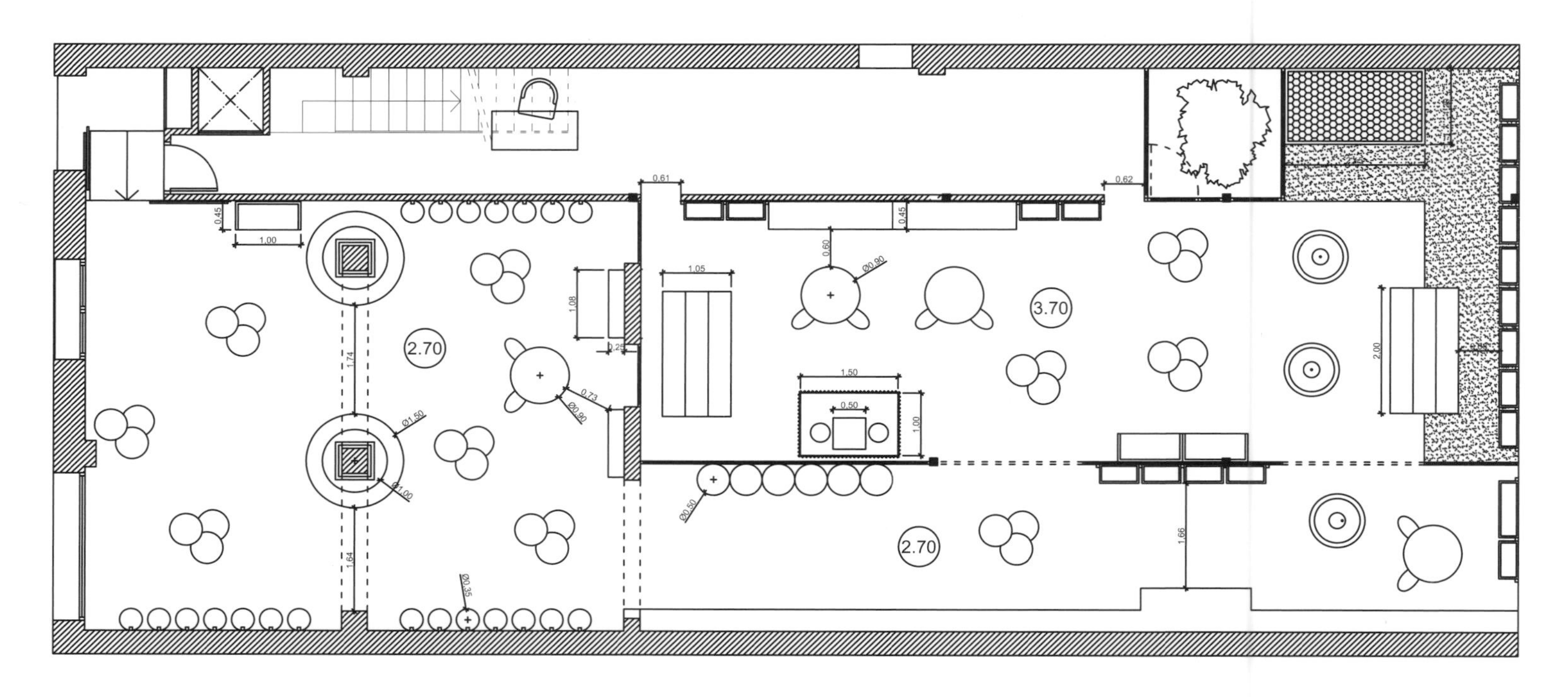
0.61
0.62
0.45
1.00
0.45
0.60
Ø0.90
1.05
1.08
3.70
2.70
1.74
1.50
0.50
1.00
2.00
0.73
Ø0.90
Ø1.50
Ø1.00
Ø0.50
2.70
1.66
1.64
Ø0.35

Farmacia Hermosa
Strepsils
NATURA MIX
Farmacia Hermosa
Farmacia Hermosa

Nutribén

SantaCruz Pharmacy

Design Agency:
MARKETING-JAZZ

Client:
Elsa Acosta

Location:
Santa Cruz de Tenerife, Spain

Photography:
Ikuo Maruyama

NUXE
Sensilis

Reafirmantes
Corporales
Promocion

Farmacia
B C N
Al Si P
Li Be
Na Mg
K Ca
Ti
Cr Mn Fe
Mo Tc
W Re
Cu Zn
Ag Cd
Au Hg

An Integrated design of a new brand concept in the pharmacy sector.

SantaCruz Pharmacy, “A healthy world”: A 300 m^2 Pharmacy organized into categories, where self-service shopping is very easy and pleasant. This is the new brand and pharmacy concept created by MARKETING-JAZZ for the Elsa Acosta Licensed Pharmacy in Santa Cruz de Tenerife.

The pharmacy is divided into areas, according to the products being sold, the desired shopping experience and the brand positioning. Special care has been taken to ensure that customer traffic flows and covers the whole pharmacy. There are waiting areas for the elderly next to the counters, where they can be served without having to get out of their seat, with the SantaCruz pharmacy staff doing their “health shop” for them.

Every product is organized into categories and, to facilitate the identification of products, lamps/bags have been designed with the name of the category, thus making it very easy for the customer to shop alone. What a surprise when you stop in front of the counter and discover that the medications are presented on a large periodic table! This immediately brings out a smile, something which definitely makes the shopping experience memorable.

It’s an area specializing in dermo-cosmetics containing a “make-up bar” where technology is used to provide the customer with an analysis of their skin and the most suitable combination of cosmetic products. An impeccable multi-brand presentation using crafted frame mouldings and natural stones reinforces the impression of a high-quality product. Farmacia SantaCruz, a healthy world.

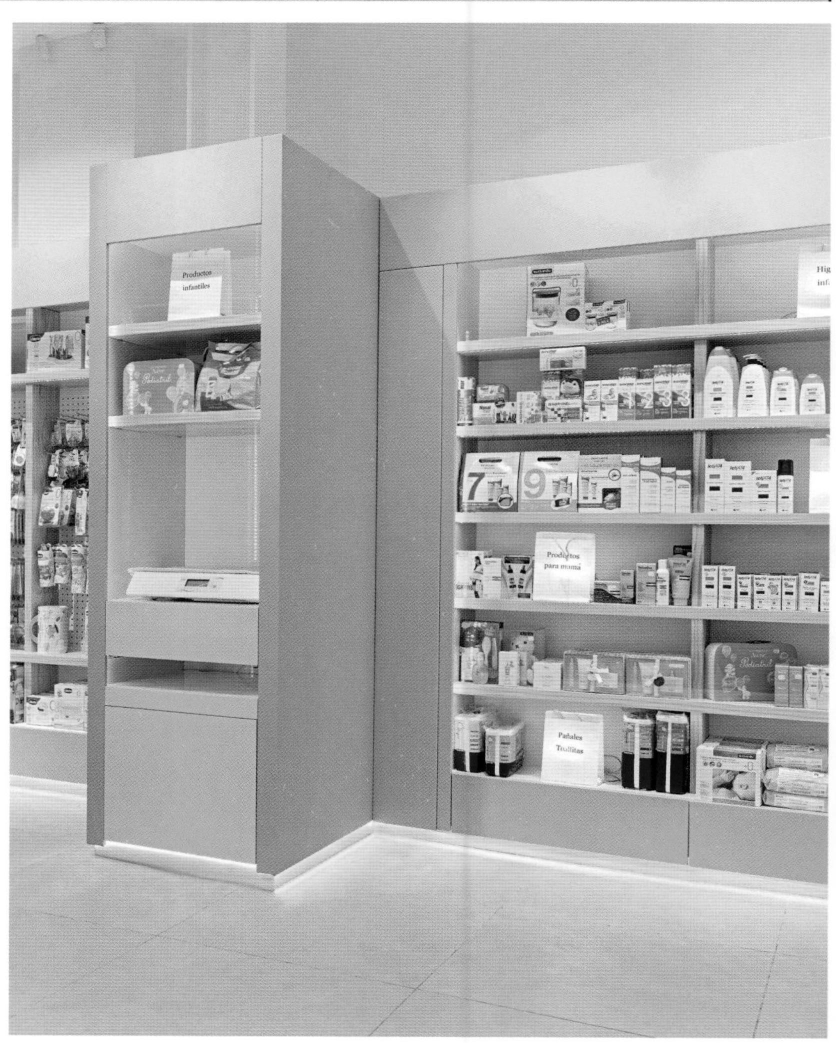

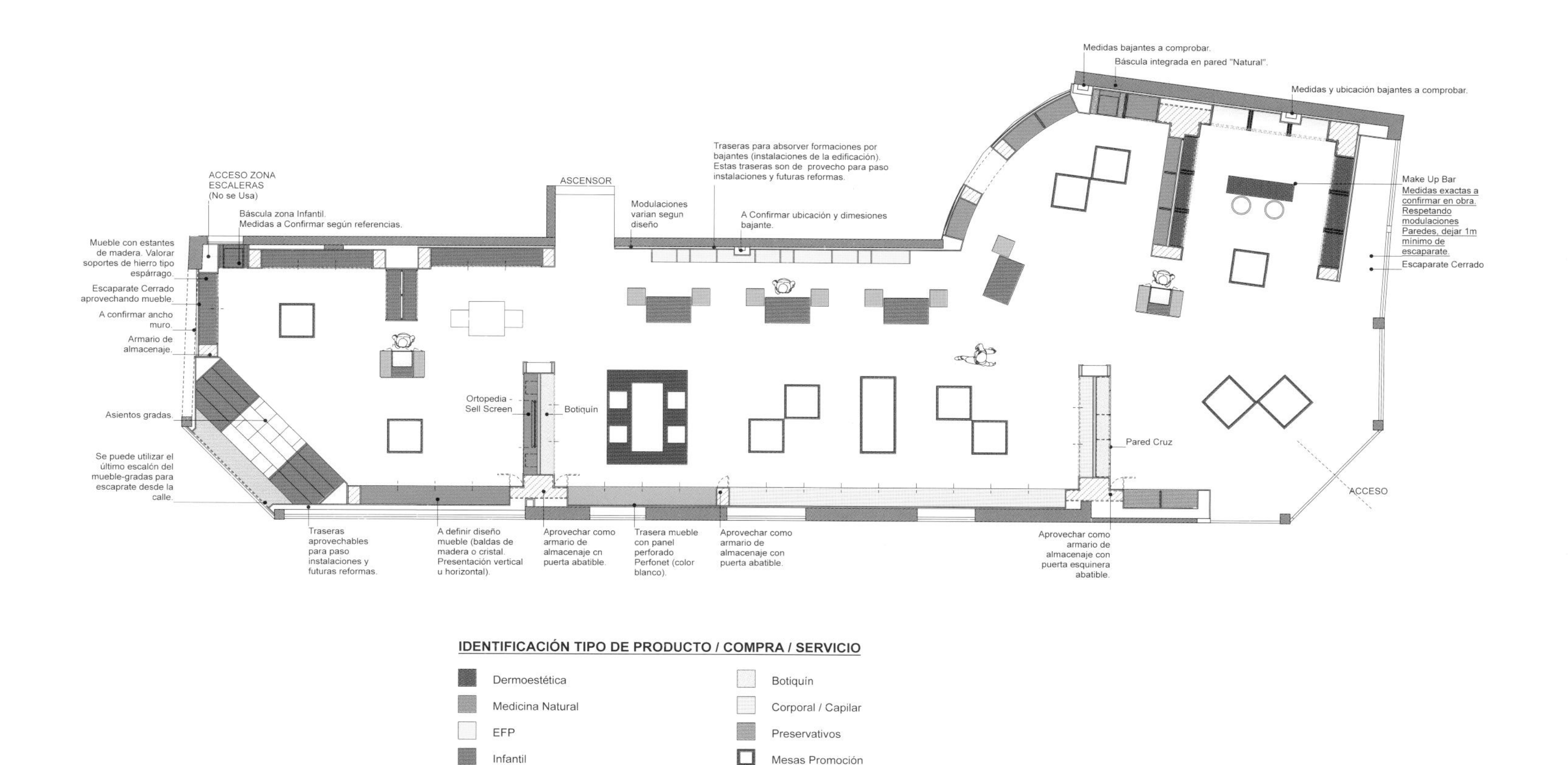

IDENTIFICACIÓN TIPO DE PRODUCTO / COMPRA / SERVICIO

- Dermoestética
- Medicina Natural
- EFP
- Infantil
- Bucal
- Ortopedia
- Dietética
- Botiquín
- Corporal / Capilar
- Preservativos
- Mesas Promoción
- Cajas
- Compra por Impulso

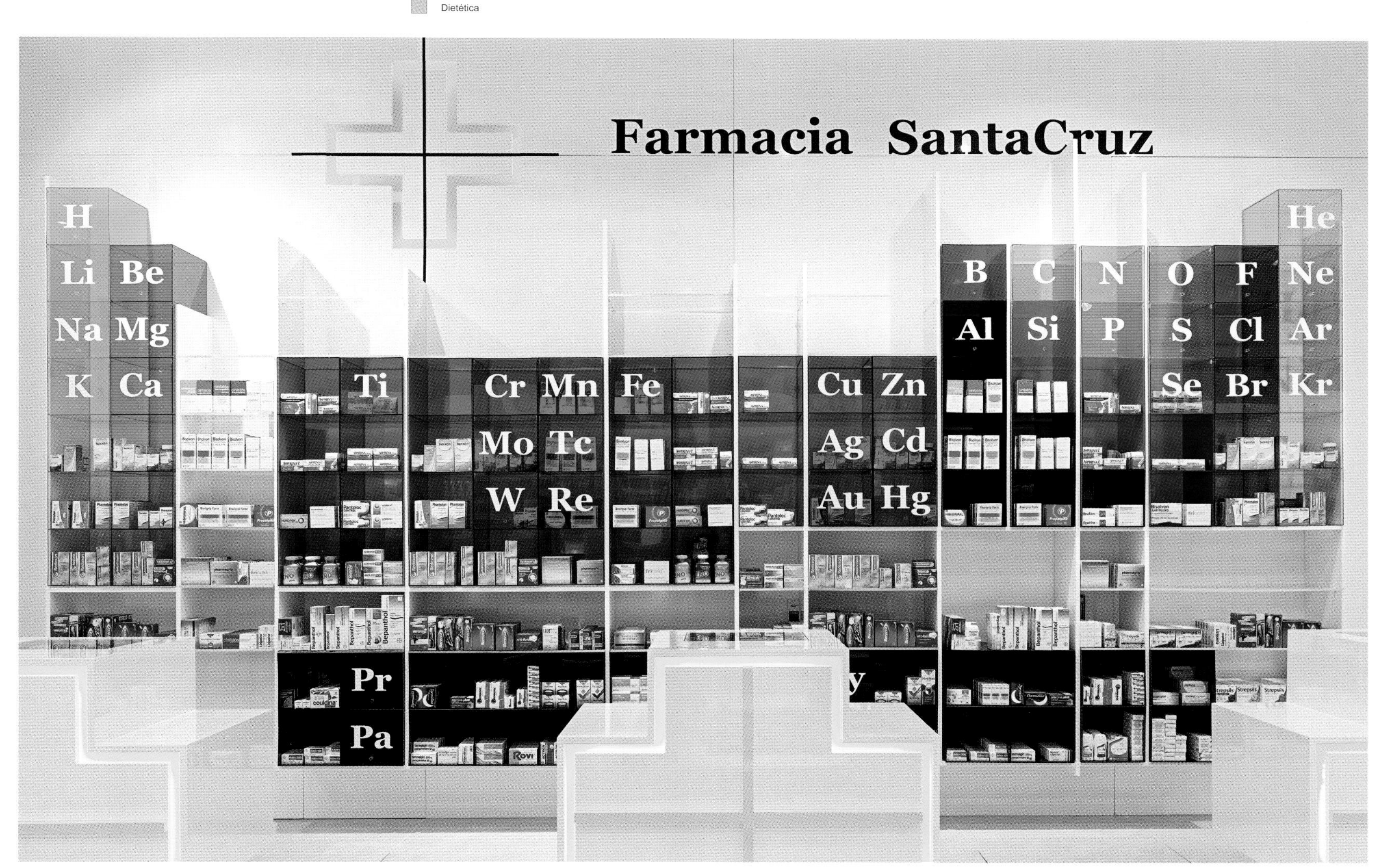

FOOD & DRINK STORES

Omonia Bakery

Design Agency:
bluarch architecture + interiors + lighting

Designer:
Antonio Di Oronzo

Location:
New York City, USA

Area:
93 m^2

Photography:
ADO

This bakery is a brand new project for the family behind the renowned Omonia brand famous for its Greek pastries. It sells pastries and breads prepared on premises in the see-through kitchen. The design of this store celebrates indulgence, the suspension of one's everyday grind through the consumption of a sweet delight. The space is soft and warm, sexy and decadent as chocolate. Much like the physiognomy of a pastry, this design wants to offer the exciting anticipation of a pastry in-fieri, the liquid concoction, and the minced ingredients. The space shifts organically with the narrative of flavors as patrons taste the succulent goods.

The main feature of the interior space is a fluid surface (clad with ¼ chocolate brown Bisazza tiles) which covers the ceiling and the side walls to different heights. This surface warps in bubbles and negotiates a system of tubular incandescent light bulbs and an arrangement of red cedar wood spheres. The epoxy flooring continues to the walls via filleted corners. A shelf and LED strips navigate the transition with the chocolate surface.

The kitchen is exhibited to the public, as it sits simply within a tempered glass box. Therefore, the exquisite level of craftsmanship of the project (with its unforgiving alignments and complex details) is paralleled with the refined artisanship of Omonia's pastries.

Diageo Concept Store

Design Agency:
Fourfoursixsix

Location:
Bangkok, Thailand

Client:
Diageo

Area:
150 m^2

BRANDY & IMPORTED WHISKY
WHITE

SPARKLING WINE
RED WINE
WHITE WINE

Fourfoursixsix won an invited competition to produce a concept store design for drinks brand Diageo. The core of the brief was to provide an innovative design that enhanced the customer purchasing experience and create a benchmark retail concept within this sector.

The design takes inspiration from luxury retail window displays that highlight key products in order to drive brand value and desirability. The plan utilises curved forms to differentiate between product types and provides smaller intimate spaces within the store.

The design creates a gallery-like space, a ribbon of display forms a linear band throughout the space, attracting attention and allowing customers to easily navigate the space and view products in an ergonomic manner. Coupled with distinctive signage, the customer is better educated about the product and more able to make an informed and enjoyable buying decision. For the client, the design was also an effective way of increasing product density in the relatively small space ensuring that revenue per square metre predictions were easily met.

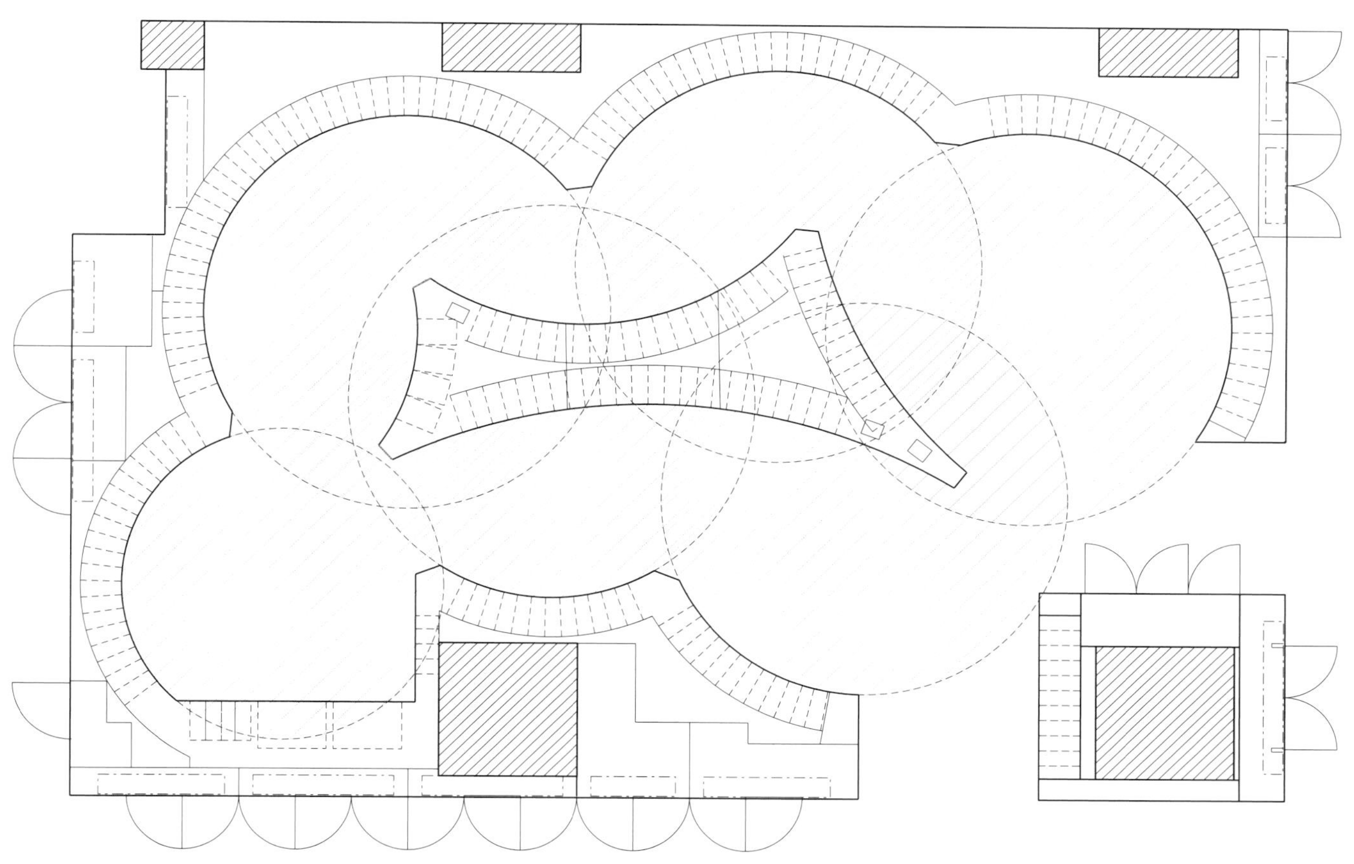

SPARKLING WINE
MOËT & CHANDON
CHANDON
10,775
5,418
4,439
4,089
4,101
3,914
1,159
990

Wein & Wahrheit

Design Agency:
Ippolito Fleitz Group

Photography:
Zooey Braun

LUSTAU
MESSIAS
PORT
TAWNY
Niepoort
TAWNY
PURO
MONTES ALPHA
COLUMBIA-CREST
Bonterra
MITOLO
WALLACE
PORTO
Ruby
TAYLOR'S
20
KAIKEN
Ultra
BRAMARE
MONTES
RAVENS
WOOD
HEARTLAND

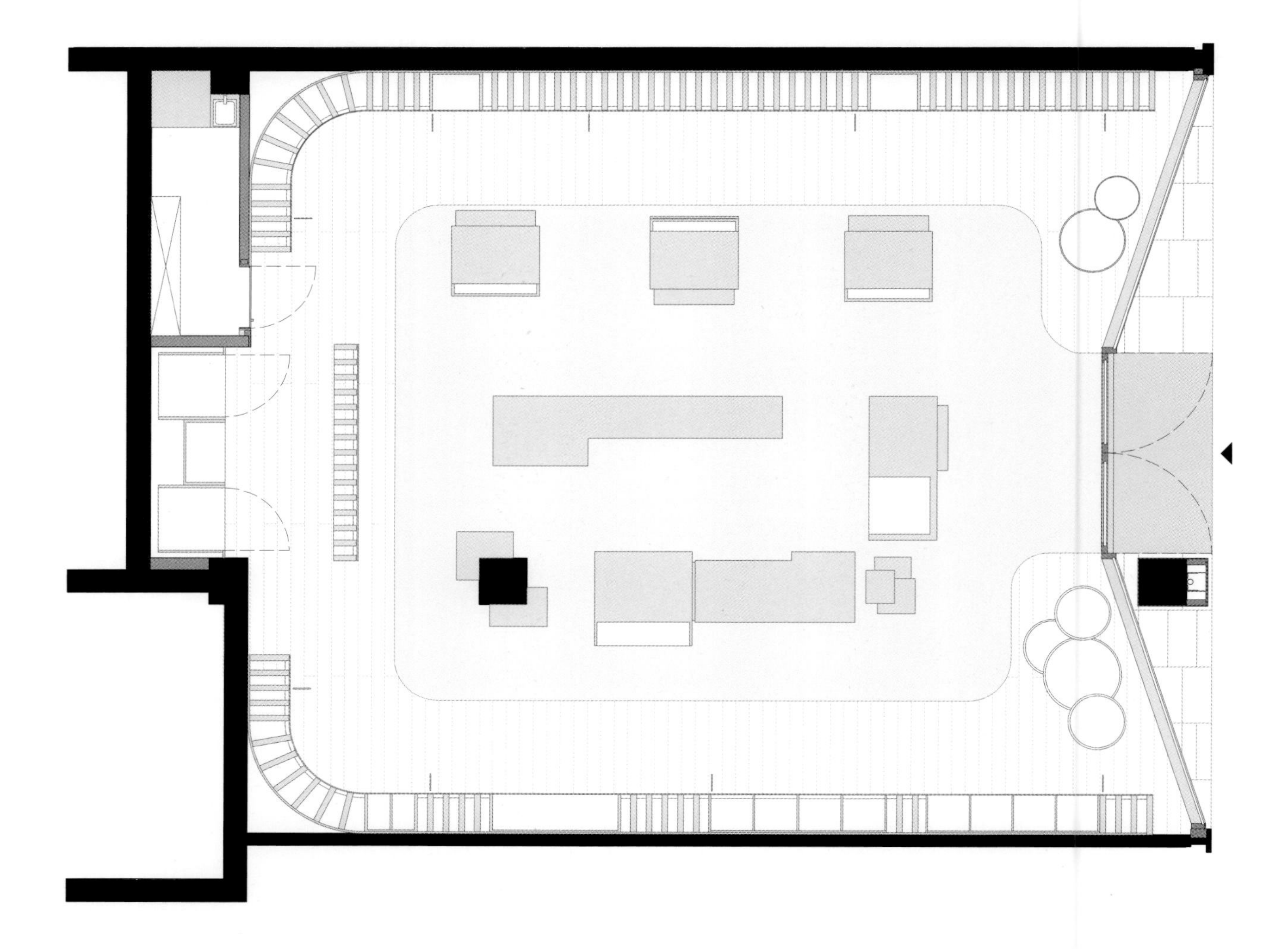

The Weinkellerei Höchst has opened a second wine shop in the new extension of the Main-Taunus Centre. The new store boasts a floor space of 80 m². The task was to present its range of over 600 different wines and spirits, delicatessen items and gifts within this compact space. The wide range of wines and the retailer's extensive wine knowledge are highlighted here in a sensual setting intended to animate passers-by to purchase. Therefore the first impression from the outside is one of undiluted wine expertise. Resembling a library of learned tomes, wine bottles fill the store from floor to ceiling along all three interior walls. A mirrored ceiling band running around the edges of the space further multiplies this effect in the vertical. Within the mirrored ceiling a canopy of glass vessels is suspended, forming a strong key visual. The light breaks in them like candlelight reflected through a glass, giving this otherwise modern setting the atmosphere of a wine cellar and conjuring up associations of epicurean indulgence.

The shop façade has been drawn into the space to optimise the available area. The doors thus open outwards so as not to block browsing space within. The tapering, funnel-shaped entrance also creates a suction effect to draw the shopper into the space. Both windows and doors are framed in black. One of the two existing pillar supports has been boxed in black and integrated into the entrance area. The name and logo of the wine shop are illuminated in white on the pillar and across the lintel, giving the façade an exclusive feel.

The two long walls are completely encased in shelving units made from white lacquered and brushed oak that curve to meet the end wall. The 501 separate compartments contained in these units are strictly organised to create a homogenous fin-like structure, broken only by square shelving elements inset in a fresh green. At the end wall stands another shelving unit with an additional 98 compartments for sparkling wines. These sit behind a continuous glass front, which is backlit to create a visual focal point at the end of the room. This unit is flanked by two chalk boards that are used to advertise special offers. A concealed area behind contains refrigerators and another shelving unit that holds high-end sparkling wines and champagnes.

The centre of the room is formed by a group consisting of a checkout counter, an elevated table for wine tasting and a packaging station. Between this grouping and the right-hand wall stand three additional presentation units. Two auxiliary vertical shelving units are fitted to a pillar. All the fittings have been given clear and sculpted forms. Their cubic geometry emphasises the material interplay of white laminate and bleached oak. All the individual store fittings are contained within a soft-cornered rectangle of epoxy resin-coated concrete executed in the floor, which stretches out a tongue toward the entrance. The flooring outside this inner zone is made of rough oak floorboards.

The contrasting floor zones are reflected in the stunning ceiling design. Here a recessed rectangle of ceiling is fitted with hidden spotlights. Within this form hangs a canopy of 150 hand-blown glass vessels in four different shades. Two thirds of them are illuminated by LED lighting. The ceiling all around is mirrored, stretching the height of the space and multiplying the already large selection of wines on display.

Glass and oak are the dominant materials in the space, both chosen to reference the world of wine — wine bottles and wine glasses, oak casks and cork oak. Their use result is a sensual ambience that appeals directly to the epicurean shopper.

A new Corporate Identity was also developed to reflect the store design. The new name of the shop, 'Wein & Wahrheit' (Wine & Truth), underscores the expert advice on wine that you can expect to receive. The elegant typography and logo design address a class of customer, who places inordinate value on quality when making a purchase.

Wein der WOCHE
ROT

Patchi Dubai Mall Shop

Design Agency:
Lautrefabrique Architects

Client:
Patchi LLC

Location:
Dubai, United Arab Emirates

Area:
380 m^2

Photography:
Luc Boegly

Located on the ground floor of the Dubai Mall, one of the largest shopping centers on the planet — 54 million visitors in 2012 — the Patchi store, designed in 2008 by lautrefabrique has been given a complete make-over to fit the image of it's new brand identity, that of a leader in the market for upscale gift chocolates in the Middle East.

In 2010, a space became vacant next to the store. The extension was made possible. Although the project was now viable there were still several difficulties. The task was to bring together two elements that were totally different both in size and in condition in order to create a single "seamless" space that was twice as large (380 m^2 in total) thus setting a new, luxurious and dramatic scene that would echo the brand's new slogan "Patchi, chocolate gifts destination".

All this was to be done remotely and in two phases, with the work being carried out at night so as not to interrupt sales (546 clients per day, opening 12 hours a day and up to 24 hours a day on public holidays). A considerable challenge with nine months of research and design and four months of construction.

Lautrefabrique has designed Patchi stores for nearly a decade, showcasing their concept : chocolate jewels decorated, dressed and presented as a jeweler would in a multitude of boxes, organized in distinct categories : chocolate cellar, new collection, wedding gifts, birth, tableware ...

While remaining faithful to the luxurious standards but new banner of the brand, the flagship shop must also relate to the shopping excursion spirit unique to the Dubai Mall where it is located opposite the doors leading on to the famous Burj Dubai Fountains.

Without redoing everything from top to bottom, it was vital to remove any hint that there had once been two stores here. In bringing together two volumes lautrefabrique designed a huge, 30 m transparent showcase punctuated by a giant black square measuring 4.5 m by 4.5 m, the famous chocolat maker's illuminated sign, flanked by two flag signs forming the entrance.

This focal point is a trick to conceal various joints and pillars and to mask the concrete structure of the shopping center.

But the unity thus created must also extend to the interior of the store where the original style has been retained and which cleverly unfolds over the walls, ceilings and floor. Curves appear, in reassuring bulkheads that soften the shelves and in a cylindrical tower (the gourmandine tower) which develops into a clever display. Nearby, in perfect harmony, a spectacular ebony case lends its deep brown curves to the brand's new boxes . Everywhere, bright RGB wells run around the cylindrical elements, enlivening them with light. A clever system of cantilever shelving with both glass and solid shelves and walls of drawers alternate horizontally and vertically while perforated partitions modulate the enlarged volume, while maintaining the advantages of transparency. Close by, between immaculate shelves and rounded walls, a peaceful lounge area welcomes the undecided guest... As for the floor, a no less essential actor of this elegant work of unification, it was covered with ultra-thin tiles (3 mm thick) allowing a clean pattern.

Black or white, ebony brown, a new version of lime green, retro-painted glass. Tens of thousands of Patchi references now enthroned in a beautiful box. As if by magic but all the while compromising with the inevitable pitfalls of working from a distance, the new shop is most definately a single entity. Everywhere one looks, the architecture, materials and colors sing the delicious notes of harmony and elegance.

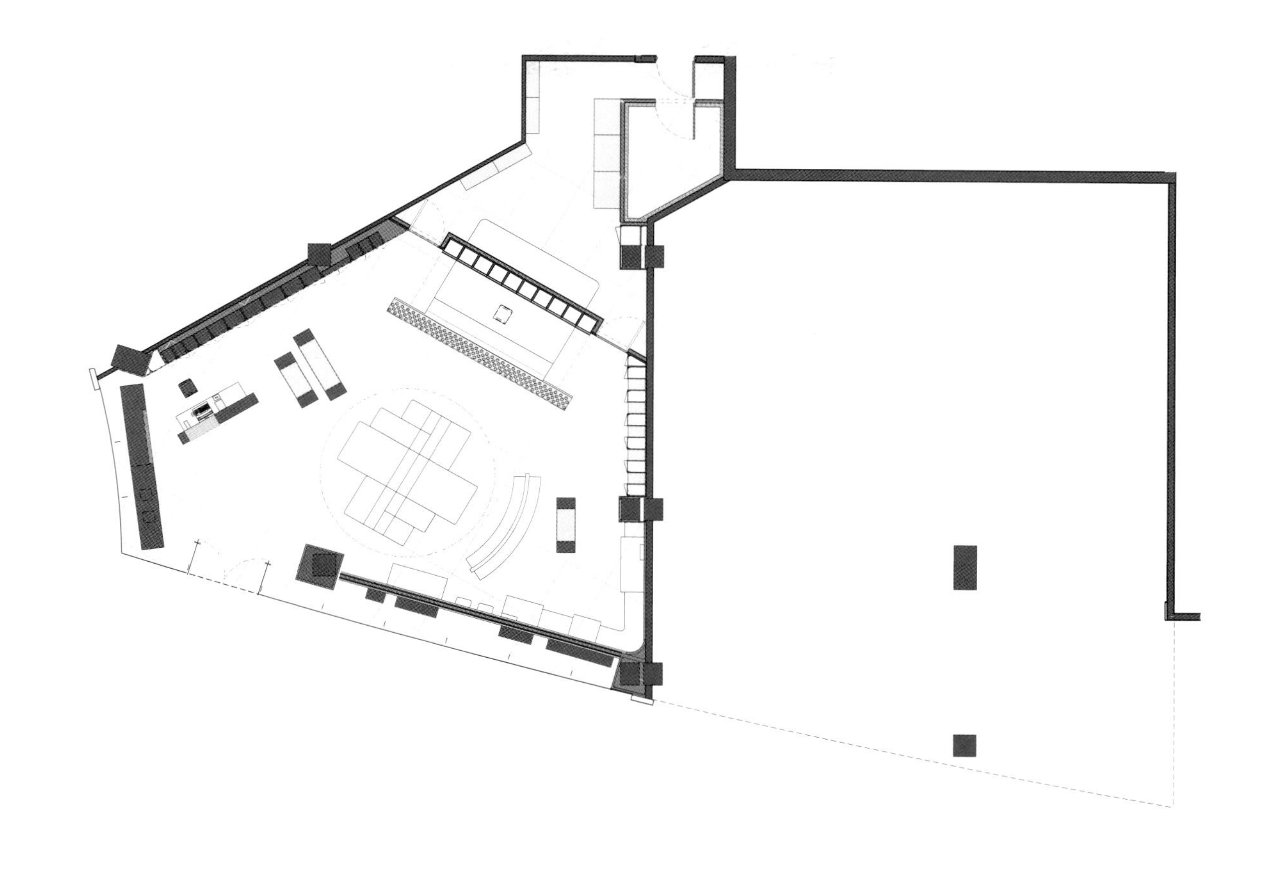

0 1 2 5 10

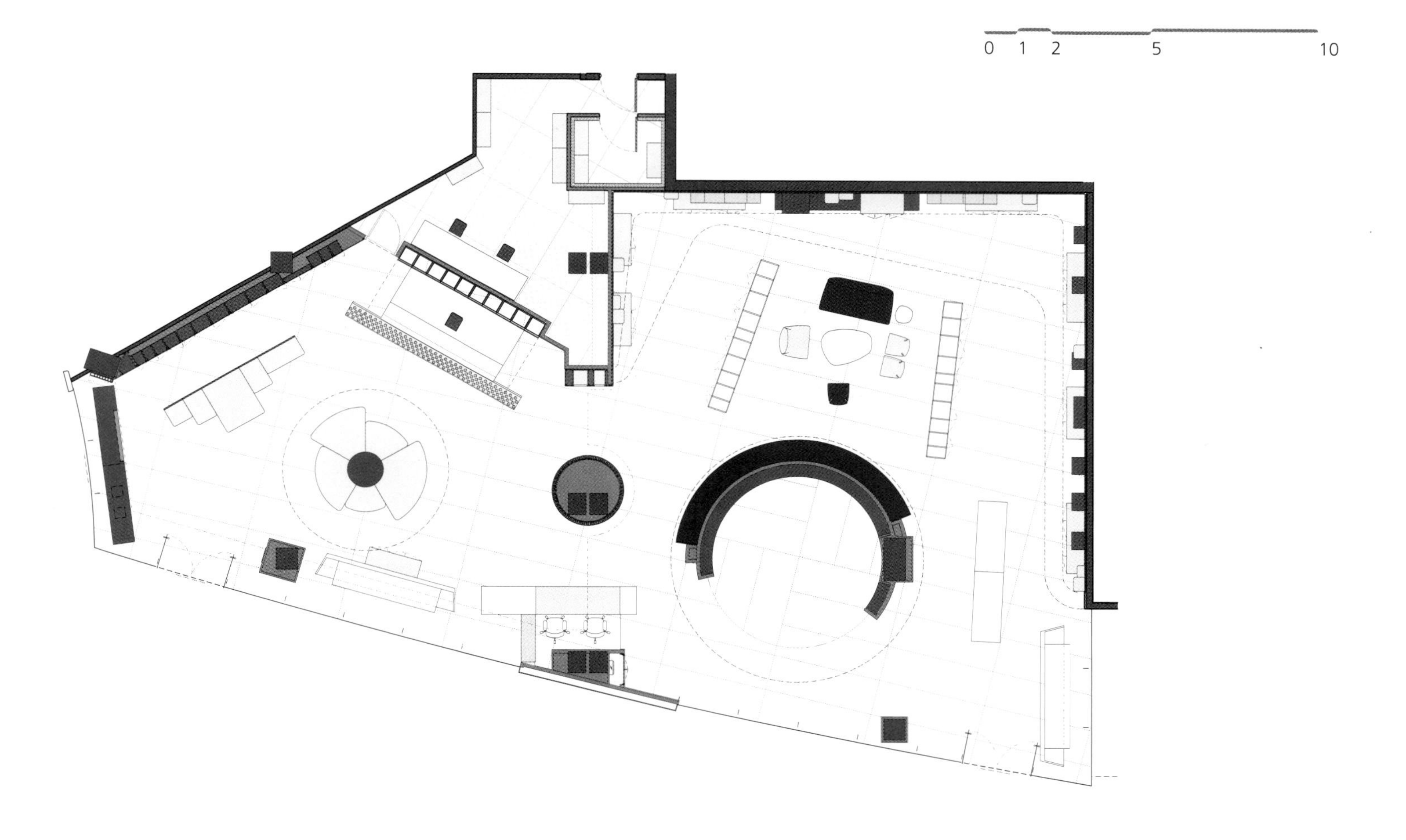

0 1 2 5 10

Patchi Souk Beirut

Design Agency:
Lautrefabrique Architects

Location:
Beirut, Lebanon

Photography:
Luc Boegly

Client:
Patchi Sal Beirut

Area:
25 m^2

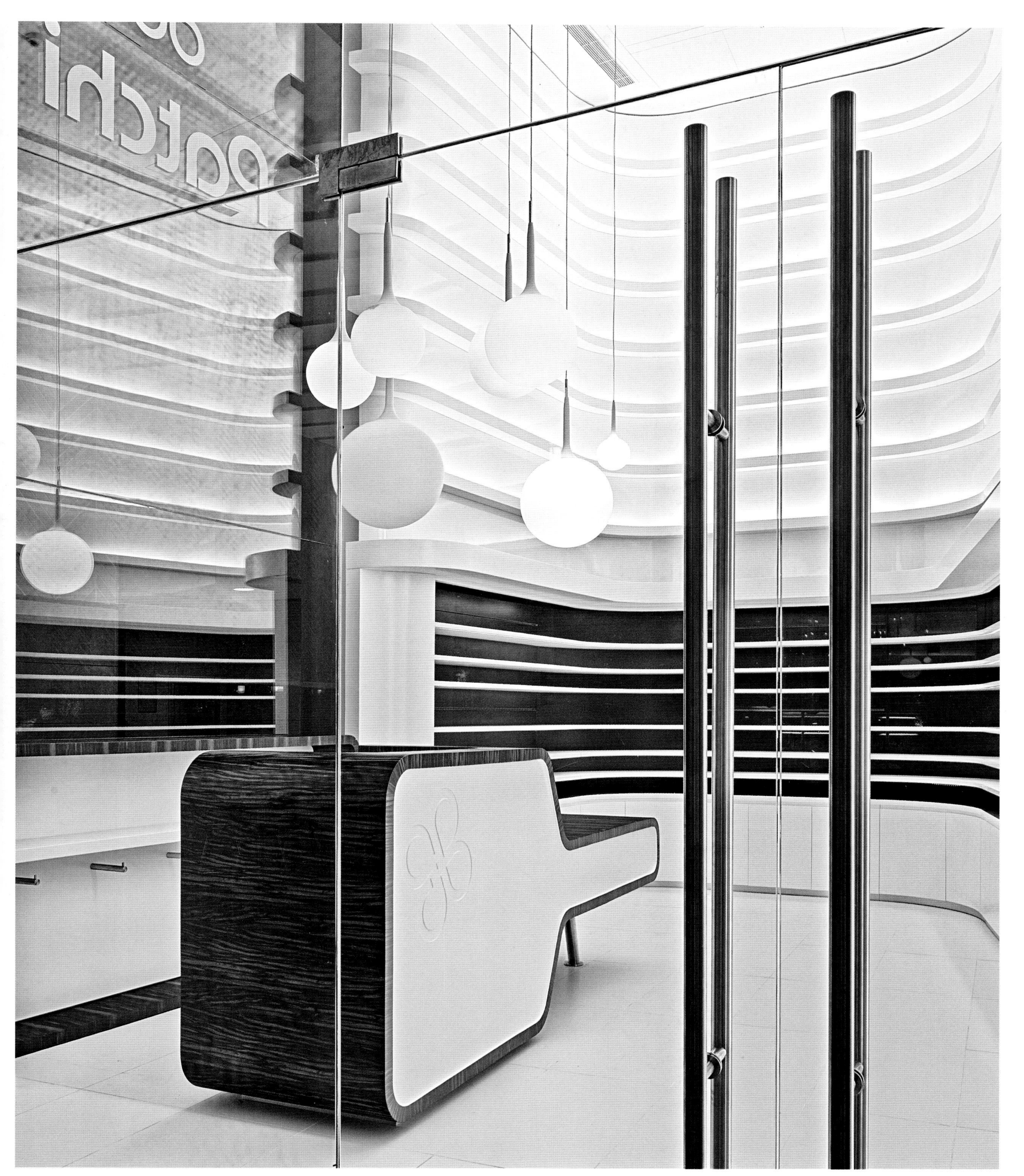
patchi

The historic souks of Beirut, located in the renovated area of Solidere, in the city center were destroyed by civil war in Lebanon. They were completely rebuilt by the Spanish architect Rafael Moneo and reopened in 2009. It is in this new commercial and tourist complex that the Lebanese chocolate maker Patchi has opened a new shop.

On this occasion Patchi has decided to break with the codes that have contributed to his reputation throughout the Middle East, with "chocolate jewels" presented on a plexiglass cone. No display stands, but only the boxes of the extensive "Gourmandine" range.

The shop is located at the corner of two indoor streets, facing a very busy square. The site covers only a small area — just over 25 m^2 — but has a comfortable height – 7m, and the creation of a mezzanine was strongly suggested by the client. However the installation of a staircase and an elevator at the rear of the shop proved to take up too much space. The architect then suggested enhancing the double-height by a scenic device favoring its expression.

The trapezoidal shape of the plan highlights the perspective and the perception of the depth of the shop transforming this cramped area into a generous space opening onto the outside.

The linear display of products is located in hollows in the walls that hug the contours of the trapezoid. Boxes and chocolate bars are displayed on cantilevered shelves that run along the whole length of the wall and whose spacing varies to accommodate all sizes of products. Linear LED’s installed under each shelf highlight the product facings. The lower walls are smooth, concealing stock cupboards.

At the top, the cantilevered shelves look like streaks on the full height. They incorporate a system of indirect lighting that allows the multiple perforations in the ceiling to be preserved.

The counter, with a Macassar ebony veneer, houses the till and the sound equipment and enables new products from the “Gourmandine” range to be presented.

To the rear, a screen, which repeats the shapes and materials of the counter, serves as storage for bags hanging from hooks made from brushed stainless steel.

Patchi Takhassussi

Design Agency:
Lautrefabrique Architects

Client:
Patchi Saudi Allied

Location:
Riyadh, Saudi Arabia

Area :
707 m^2

Photography:
Luc Boegly

Patchi

Patchi
Patchi
Patchi

Lautrefabrique has designed stores for Patchi, a leader in the luxury gift chocolate market in the Middle East, every year for the last ten years. In 2000, in Saudi Arabia, Patchi took over a neoclassical villa on Takhassussi Avenue, a ten lane expressway in a residential district of Riyadh. In 2008, the group decided to transform this unconventional boutique and establish the reputation of the brand in this country in the twenty-first century.

The expertise of Lautrefabrique brought this ultra-typical location a new identity. Without a formal response but, as always, with an array of architectural solutions to counter specific problems: and here the agency was even more involved, taking responsability for going out to tender.

The first task concerned the external appearance of the premises. Removing the aspect of the neo-classical villa in favor of an eye-catching facade. The idea was to create a huge screen, with a focal point able to attract the attention of the many motorists who use this main road every day. White aluminum composite panels were chosen with a layout of meticulous, graphic, luminous monograms. Two small, horizontal display windows frame, in the manner of two screens, a double entrance made necessary because of the sandstorms that frequently hit of the city.

Inside, the feeling is spectacular. Huge, with double height ceilings, the luxurious, contemporary boutique of over 700 m^2 has been transformed into a glorious temple : to chocolate elevated to the status of precious gift, an offering for all celebrations of life summarized by the new brand slogan "Patchi: chocolate gifts destination".

The decor is sumptuous. With ebony, Corian, black and white retro-painted glass, walls of drawers overhung with mirrors and other transparent shelf partitions, numerous displays, the chocolate cellar, white floors and a new anis green ... The house livery colours have been respected but somewhat livened up with ingenuity and sophistication. A polished stainless steel mirror covers a pillar, creating an alluring point of reflection to the decor.

Oak makes a remarkable appearance in the surprising latticework of a long sculptured element with delicately ridged edges. It crosses the heart of the retail area on the ground floor then seems to spring out of the first floor in a sort of lattice canopy shielding a more intimate area.

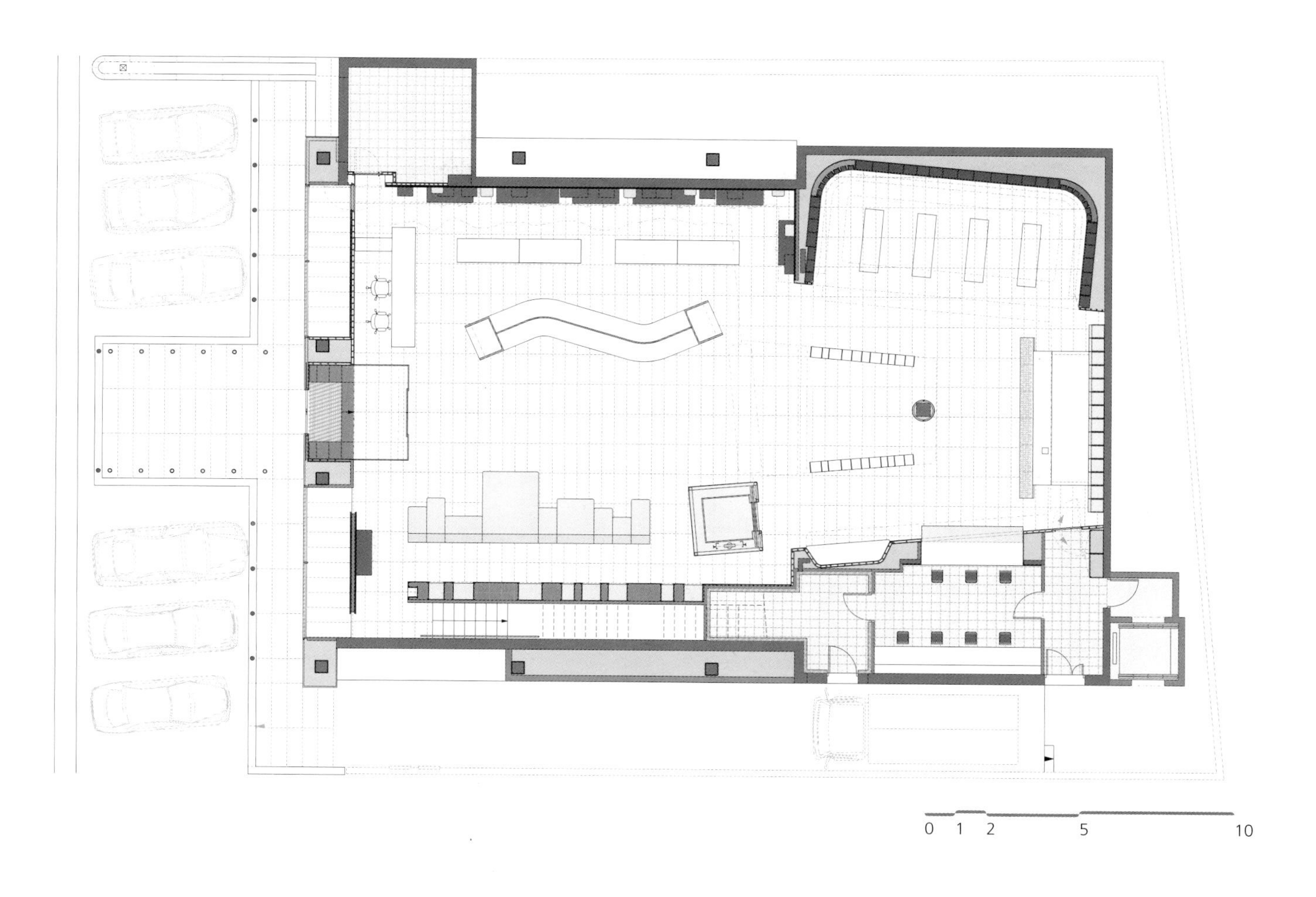
0 1 2 5 10

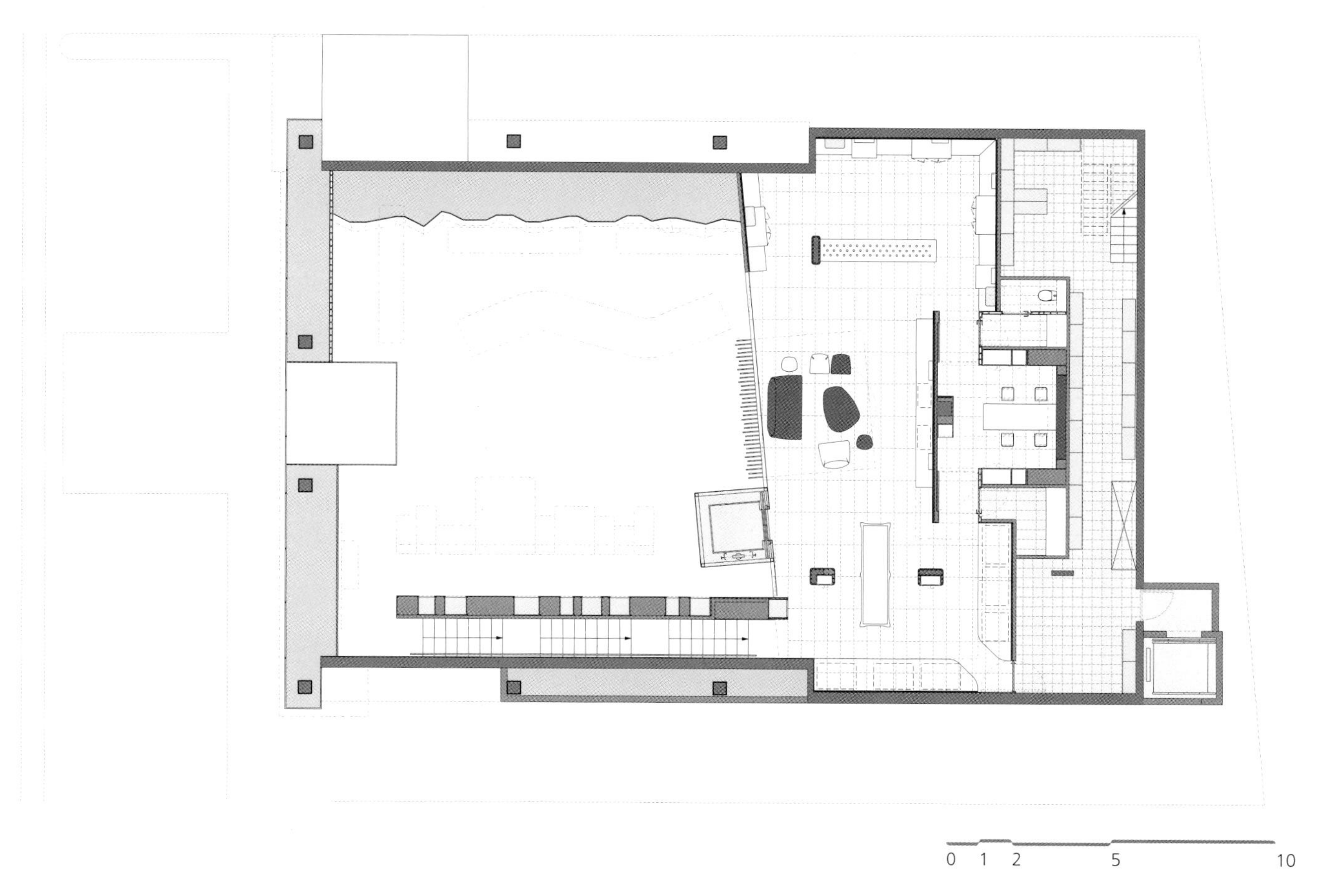
0 1 2 5 10

Blacks here, whites there, ceiling volumes competing with bulkheads, light wells and other relief. Any number of variations such as the "choco-choco" strip combines to punctuate the space to create separate areas. But there are also many tricks hiding technical elements, starting with the countless air conditioning ducts, unsightly but so essential in Riyadh where differences in temperature are extreme.

On the walls, huge transparent shelves lined with colorful pieces of crystal alternate with the depth of ebony running along displays profusely covered with bowls, trays and other containers with flamboyant delicacies. Everywhere, vertical and horizontal lines cover the space, providing tens of thousands of Patchi references with the aspect and character of a sumptuous setting.

Completely reorganised and open-planed, the store houses a panoramic elevator. A staircase comes up to join the mezzanine with the comfortable pace of a succession of carefully designed steps and softly-lit half-landings reflected in a multitude of square mirrors. The more pastel, more intimate world of the first floor is dedicated to celebration gifts. (Baby and wedding)

From top to bottom, the whole presentation of the sales area (463 m^2) pays the greatest attention to every detail by means of repetition, symmetry and cleverly orchestrated opposition... But behind the scenes in the 118 m^2 area given over to workshops and services (plus 125 m^2 of attic) the work has also been substantial, especially with the creation of an elevator that serves three levels and a staircase connecting the stock on the first floor to the warehouse located in the attic.

While remaining faithful to the Patchi image but bolder in its proposals the Riyadh store required extensive studies. Completed before Ramadan 2012, its construction lasted seven months and was carried out in two phases in order to allow the store to remain open during construction.

CHOCOCHOCOCHOCO

Winescape Wine Shop

Design Agency:
One Plus Partnership Limited

Design Team:
Mr. Law Ling Kit & Ms. Virginia Lung

Location:
Wanchai, Hong Kong, China

The wine shop is inspired by Cape Town, South Africa. The primitive-yet-modern character of South Africa is shown with the modern and comfortable style of the shop.

Every single rack of the wine shop is custom made with the shape of many mounting racks. Every mounting rack is forming several degrees of acute angle with the wall, provoking realm of illusion within the incredible tiny space. Big pendant light featuring scenic photographs of Cape Town, pattern setting of the flooring resembling African tribes and stainless steel bar and kickboard are bringing the joyful smile of South African in the interior.

Cioccolato

Design Agency:
SAVVY STUDIO

Client:
Cioccolato

Location:
San Pedro Garza García, NL, México

CIOCCOLATO
(BAKE & DECOR)
SUGAR
FUDGE
FLOUR

SUGAR
FUDGE
4
CUP
CAKES
4
CUP
CAKES

CIOCCOLATO
(BAKE & DECOR)
LIFE IS

Cioccolato is a pastry boutique specialized in custom deserts for special events. In the last couple of years, auteur pastries has grown considerably, which is why brand and product differentiation have become crucial factors for the success of a business.

The main concept is derived from the already existing Cioccolato, and repositions the brand as a pastry and specialty services provider that caters to all occasions. The Bake & Decor descriptive is used to communicate the new attributes of the company's work towards the rebranding project, without confusing Cioccolato´s current customer base.

SAVVY STUDIO developed a sweet and festive visual identity that uses brightly colored elements and memorable phrases, which go well with any kind of special event.

HOUSEHOLD STORES

Acera Flagship Store

Design Agency:
Hangar Design Group

Location:
Taipei City 106, Taiwan, China

Client:
Acera

Photography:
Hangar Design Group

Acera is a family company based in Taiwan and Shanghai and operating throughout the whole of China. Stemmed from the centuries-old tradition of Taiwanese ceramics, Acera is specialized in home accessories, offering a wide range of one-off pieces as well as a rich catalogue of traditionally decorated mugs.

For its first store in the heart of Taipei, Acera has teamed up with the Italy-based design agency, Hangar Design Group. Clothed in golden leaves and warm colours, the two-storey boutique stands out for its mix of materials: wood, gold leaf, and ceramics are used throughout the store, in patterns echoing traditional Chinese designs in a contemporary interpretation. The final look matches the brand's style based on colour, tactile sensations and decorations at a crossroads of tradition and modernity.

Black and White

Design Agency:
ALBUS Design

Designer:
Felipe Rijo

Architect:
Henrique Steyer

Classic and always in fashion, black and white dominate the winter/fall decoration for the store Gobbi Novelle, in Porto Alegre, Brazil. The new visual identity was created by architect Henrique Steyer and designer Felipe Rijo, from ALBUS Design, who used as inspiration and basis for the concept the quote by Aristotle Onassis: "I dye my hair white for business meetings, and black for romantic outings."

The aesthetic proposal was applied in the entire store and website. The "Black and White" concept begins on the entryway stairs, half-painted on each color. The facade follows the same paint scheme. The showroom has been divided as well — creating two clearly distinct settings.

On the inside, the highlights are two "libraries", one totally black, and the other totally white. With such strongly defined colors, it's like the store assumed a split personality.

Besides the black-white dichotomy, the decor includes graffiti look-a-likes in the walls and in the storefront windows, signaling a rebel attitude. Adding to this posture, light fixtures hanging from the ceiling in different heights, a mirror ball and images of burlesque dancers. Modern and vintage at the same time, these scenic features pay homage to the decadence of the night and the style of the Folies Bergère, the Parisian club where great artists performed on the first half of the 20th century.

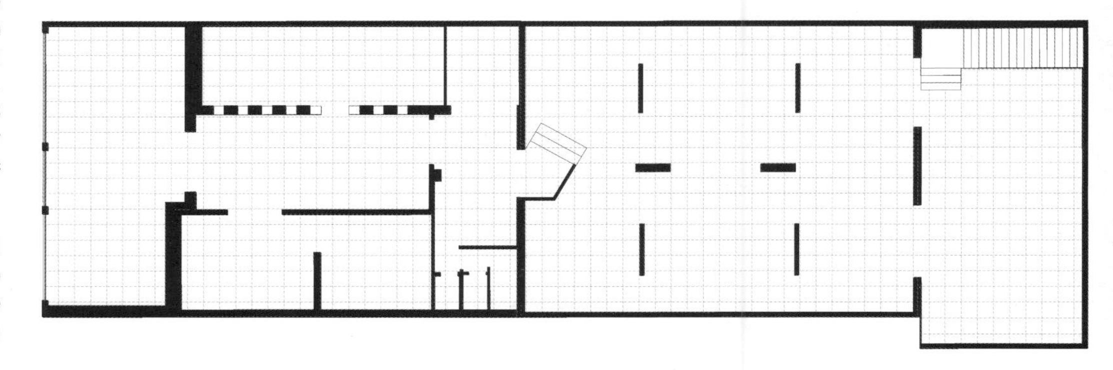

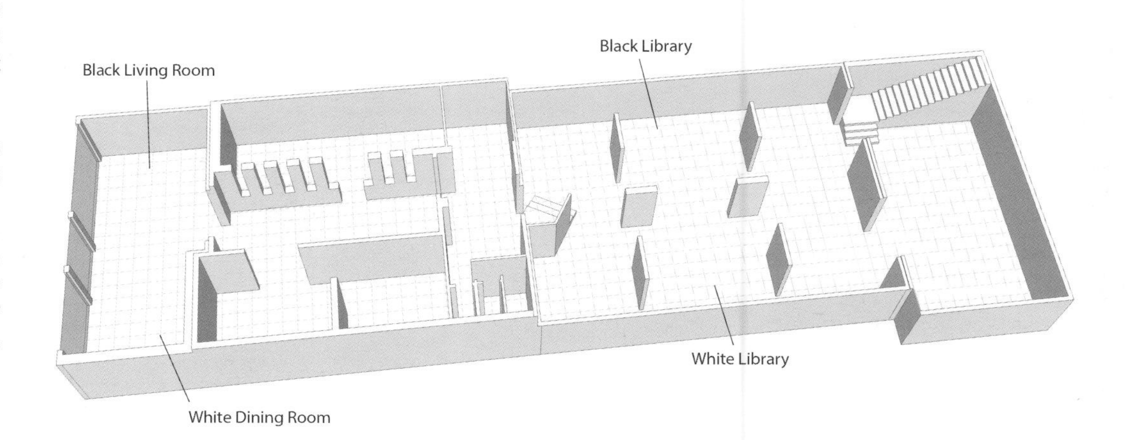

Aura

Design Agency:
Dalziel+Pow

Area:
850 m^2

Location:
Riyadh, Saudi Arabia

Dalziel+Pow has developed and designed a new fashion homewares brand, Aura, for the Middle Eastern market. Their brief was to create a brand identity, store environment and brand communications, including their packaging, to encapsulate a new confident and younger Saudi Arabian and wider Middle Eastern style.

They have created a brand that although largely "western" in influence appeals to the young style conscious Saudi consumer. For the new logo mark, logotype and all in-store communications, they designed English and Arabic versions. The brand has been built around a determination for the best quality and design in every detail, from the translucent carrier bags with bespoke opaque patterns, to the rich dark cerise backlit glass cash desk.

The store is also zoned in departments, from bedroom to bathroom, living to dining, lighting to rugs and accessories. The atmosphere of the environment is dark and moody with strong accent lighting picking out the product to a jewel like quality. The merchandise is grouped in lifestyle displays bringing together vibrant colors and patterns in the contemporary area, mixed with a more subdued palette in the classic display.

This new brand has tried to encapsulate a new confident Saudi style by bringing together products from across the globe and creating a unique mix that will excite the largely conservative Saudi customer. The highlight of the store and the Jewel in the crown is the "Iconic" area, which sits proudly in the centre of the store as a clean white glowing glass structure radiating light and pulling the customer towards it like a beacon. The Iconic area represents the heart of the brand in which traditional Middle Eastern products are restyled with a contemporary twist, making these pieces unique and desirable as never before.

H&M Home
"Home Reflections"

Design Agency:
UXUS

Client:
H&M Hennes & Mauritz AB (H&M)

Location:
Stockholm, Sweden

Photography:
Dim Balsem

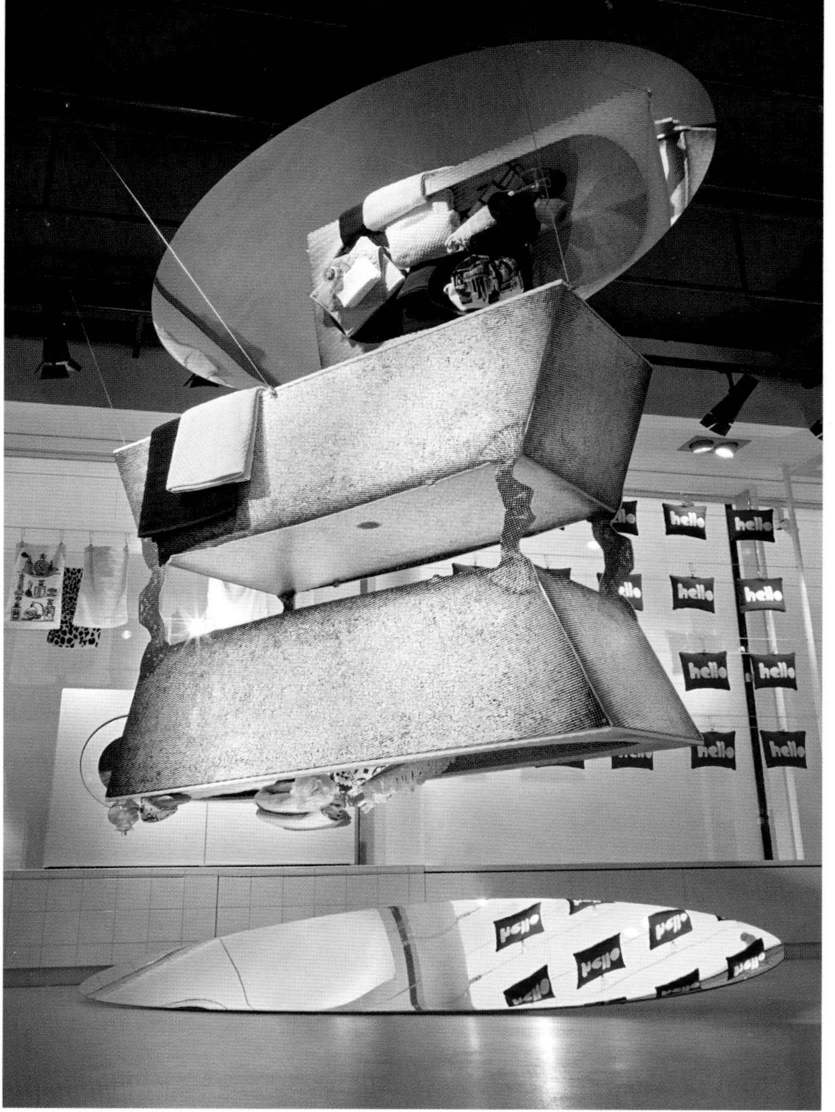

UXUS was invited to create the launch for the H&M Home collection in Stockholm. The approach was to create a "gallery" of fun fashion home products that customers are encouraged to touch and explore as they create their own unique home style.

H&M Home is a gallery/showroom using highly emotional product presentations, verging on art that encourages customers to engage with the brand. For the launch, UXUS created a display of mirrors and suspended furniture to showcase the variety of looks that can be created with the H&M Home collection to reflect each consumer's personal style.

Located in the H&M headquarters in Stockholm, the installation "Home Reflections" imagines the world through a looking glass, to explore our ever-changing relationship between identity and style.

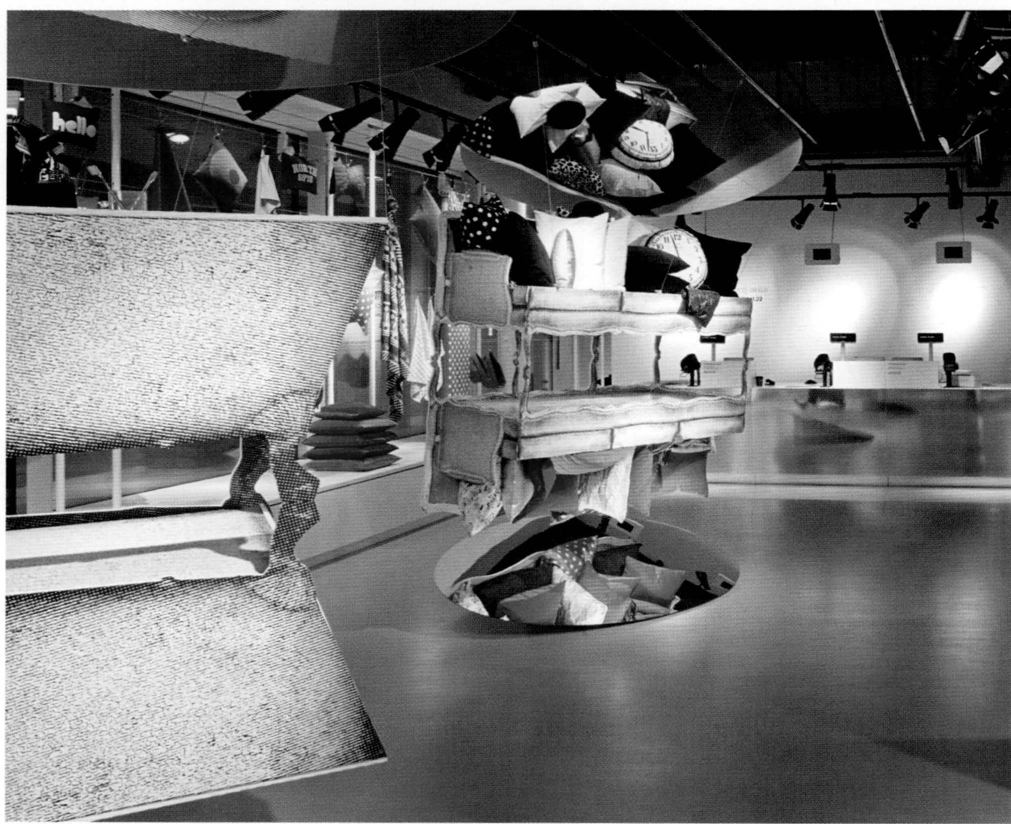

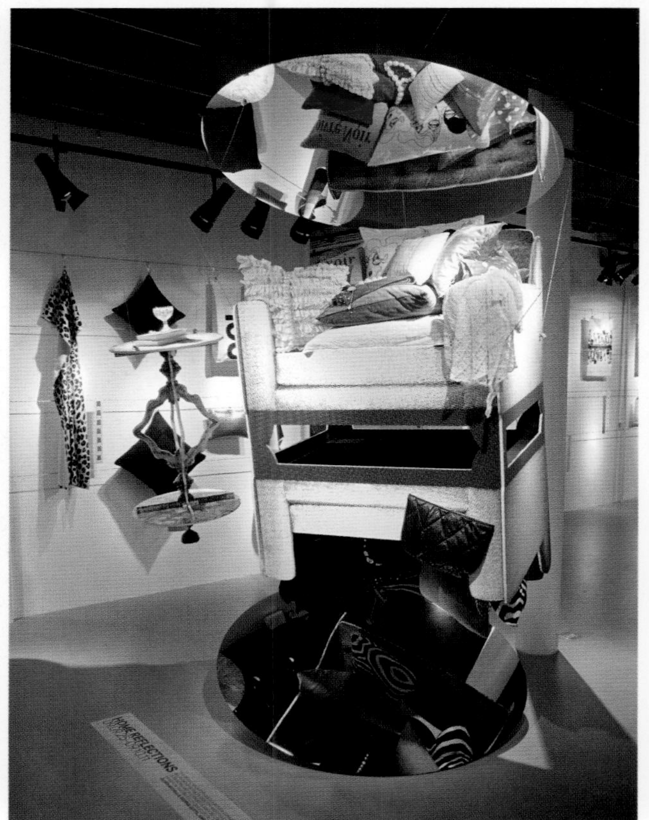

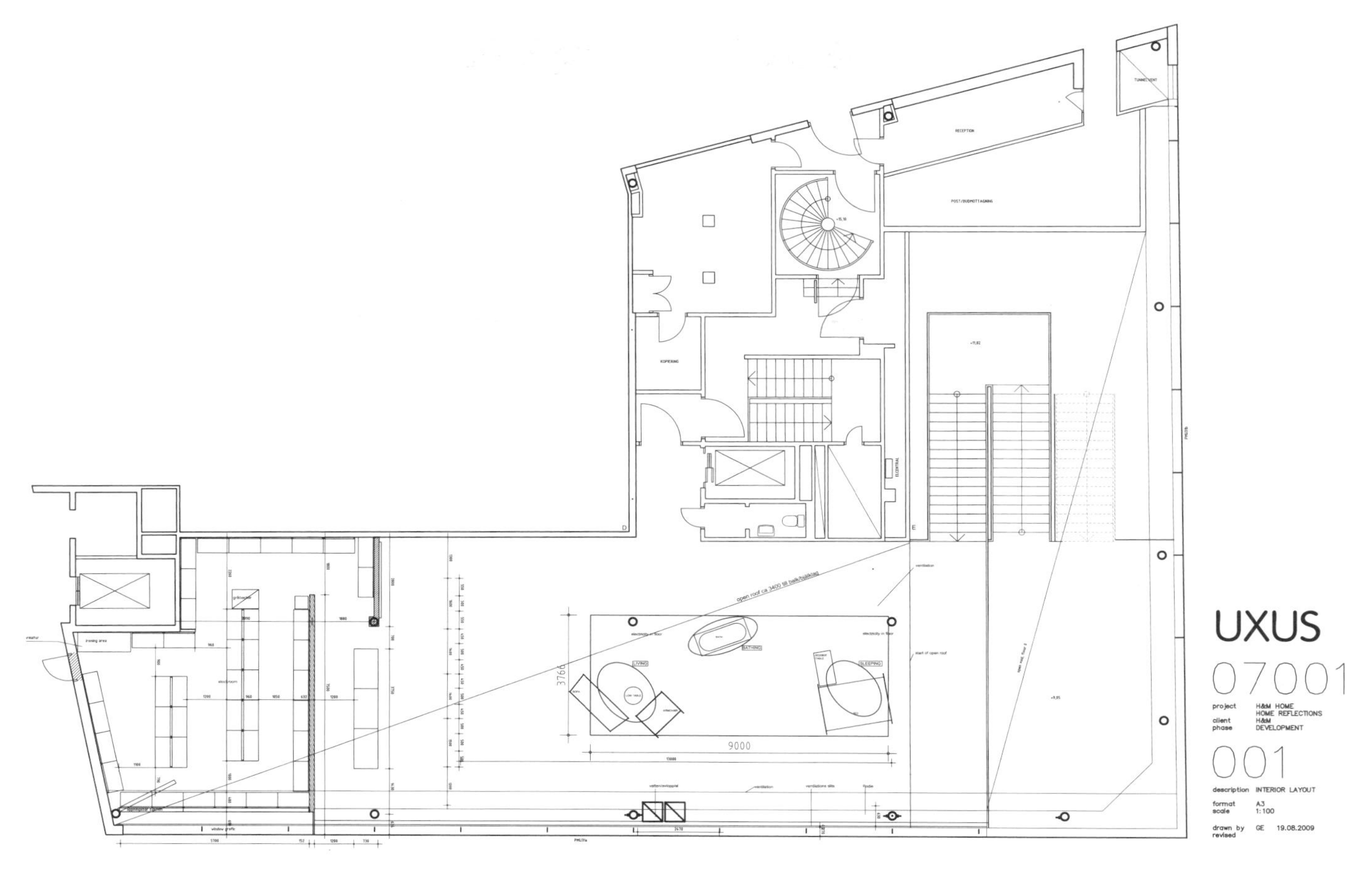
UXUS
07001
project H&M HOME
HOME REFLECTIONS
client H&M
phase DEVELOPMENT
001
description INTERIOR LAYOUT
format A3
scale 1:100
drawn by GE 19.08.2009
revised
9000
3766

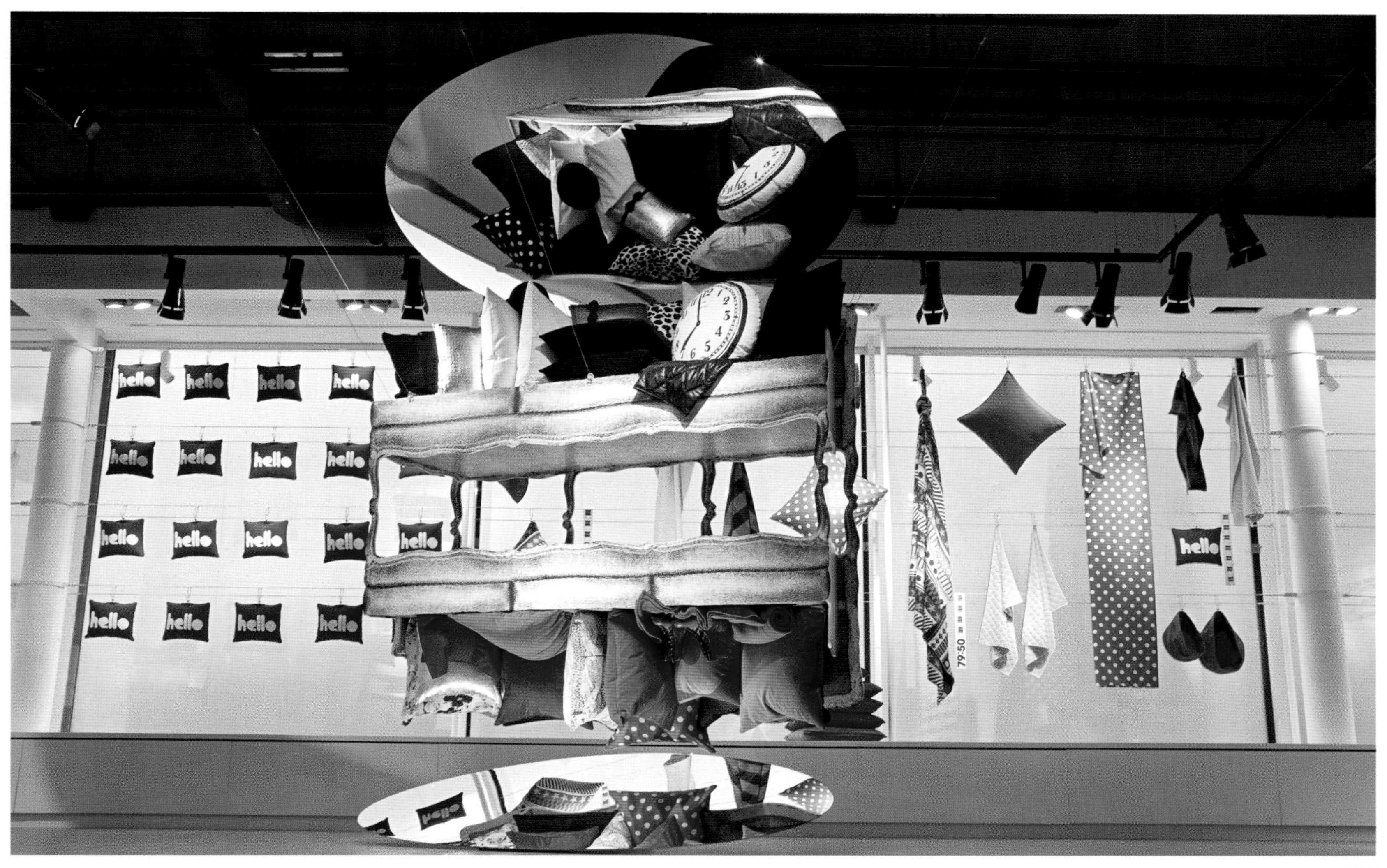
hello
79:50

JEWELRY STORES

House of Pertijs

Design Agency:
Concrete Architectural Associates

Client:
Pertijs Juweliers

Location:
Breda, the Netherlands

Area:
310 m^2

Photography:
ewout huibers for concrete

Gobbi Novelle

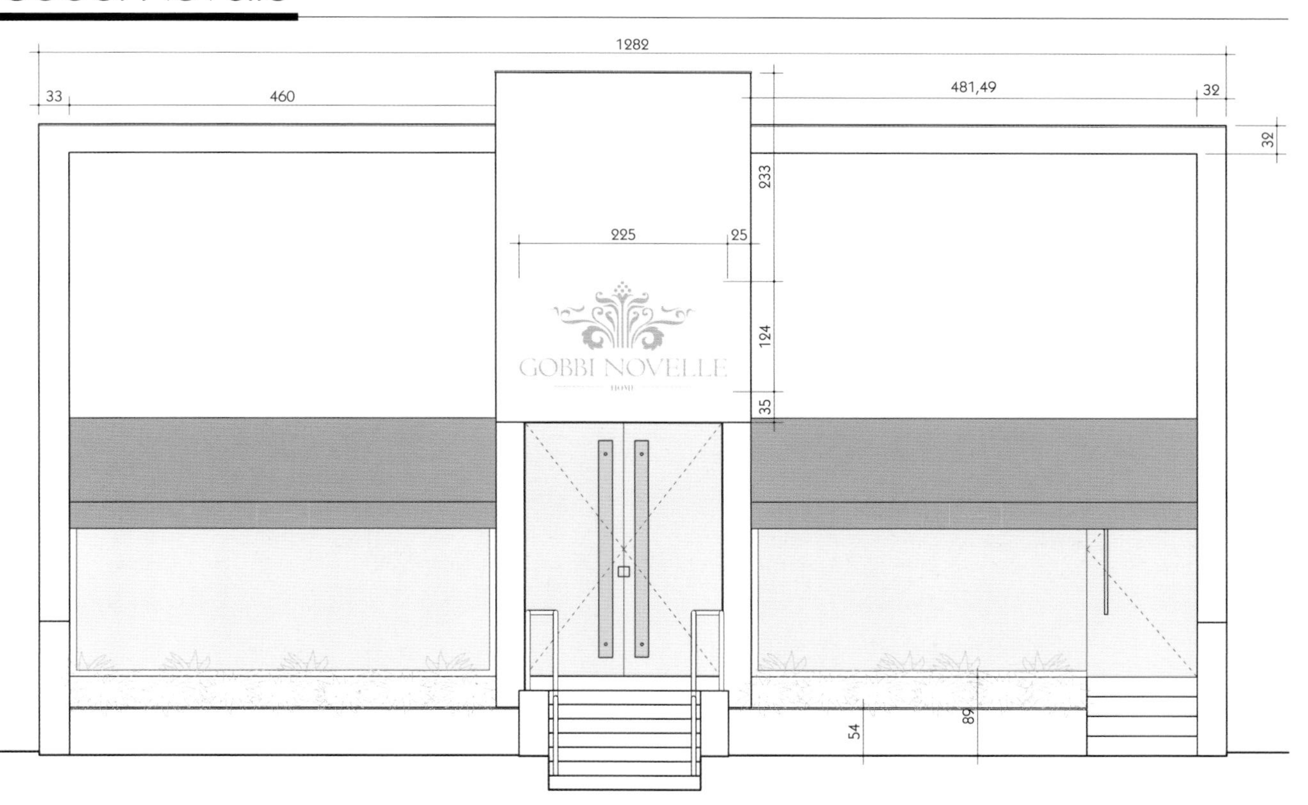

"House of Pertijs" has its roots in a traditional family-owned watch and jewellery store in the Dutch city of Breda, where clients were treated more like friends than customers. To create more customer flow, the owner of the store came up with a new vision: nowadays luxury is not merely about watches and jewellery, but also about lifestyle and its related high-end products. Why not combine these different product groups in one store? The idea of a watch and jewellery concept store was born.

The design concept of the House of Pertijs was to build a store offering a total luxury and lifestyle experience combined with the welcoming and warm feeling you get from a home. Feeling at home enhances the owner's philosophy to treat customers like friends and therefore the metaphor of a traditional 'house shape' was used as the starting point for the interior design. Pure white surfaces create the image of being inside a house in an abstract way. Not only the angular ceiling but also typical functions give the customer the sense of visiting friends at home. Sales talks can take place at the freestanding kitchen, on the sofa in the living area or in the more intimate private "dining" room.

The second objective of the design concept was to ensure that the watches and jewellery remain the customer's main focus. In former times, gold was associated with luxury. Many things have changed since then, but this old metaphor still applies. By creating display cabinets in the form of an oversized gold bar, the familiar atmosphere and more traditional appearance of the old watch and jewellery store could be recreated in the new loft-like retail space.

The gold bar is a unique piece of furniture, drawing attention to the products (watches and jewellery) showcased inside. Besides its quality as showstopper, when the store is closed the gold bar acts as a "safe". Nine individual display cabinets can be pushed up against each other and locked. By creating this "safe", most of the products can remain in the display cabinet, as there is no longer any need to deposit them in the regular safe.

OhmyGOd

Design Agency:
MARKETING-JAZZ

Photography:
Ikuo Maruyama

Client:
Catalina Hermoso

OhmyGOd, a new concept in jewelry aimed at creating a trend in the world of fashion.

The international expression and name of the line "OhmyGOd" has acted as internal for the design and decision making. One of the objectives was to visually take the store out on the street, making customers focus their attention and clearly see each of the point in sales area. The designers designed the floor, furniture, lights and graphics.

Various improvements have been made since the design of the first OhmyGOd, the flagship store at 70 C/Serrano. They are all aimed at clearly conveying the image and positioning of the fashion brand. Its location opposite to H&M and Zara rather than in the jewellery quarter, the creation of a large framed window display which makes it possible to create and develop visual communication campaigns which are more in tune with the world of fashion, the creation of a dressing room area which is dominated by a large full-length mirror to improve the shopping experience, and improved lighting to present the product.

OhmyGOd

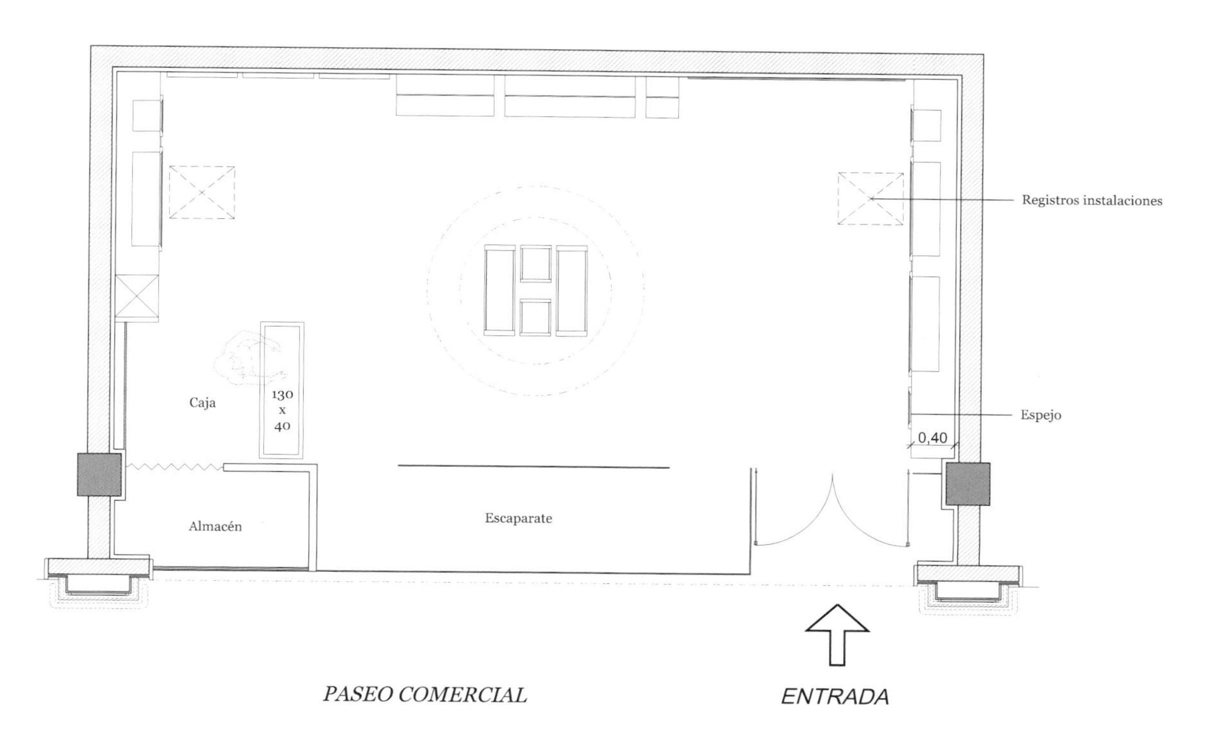
Registros instalaciones
Caja
130
x
40
Espejo
0,40
Almacén
Escaparate
PASEO COMERCIAL
ENTRADA

GOd

Burma

Design Agency:
Atelier du Pont

Location:
Paris, France

Photography:
Philippe Garcia

Client:
Burma

Area:
60 m²

RUE
LA PAIX

The front of the Burma boutique in Rue de la Paix is decked out in black and glass. The exterior and interior spaces attract and interact via a facade that takes its inspiration from the portrait gallery. The jewellery in the shop window is exhibited in hanging elliptical display cases that almost seem to float in the air; the back of the cases is reminiscent of a hand-held mirror. This display provides a bold and unusual introduction to the store.

The interior is designed like a large-scale jewellery case, as the diffuse light cast by an enormous chain-mail chandelier shimmers over the surfaces. The agency designed all the tables with moving mirrors and other furniture except the chairs. Everything is made to measure, with the greatest attention paid to ensuring that this new attire fits the boutique like a glove.

The atmosphere is muffled, the seats are comfortable and the interior styling assured. This all combines to form a refined interplay of straight lines and ellipses that creates a fascinating and magnetic environment.

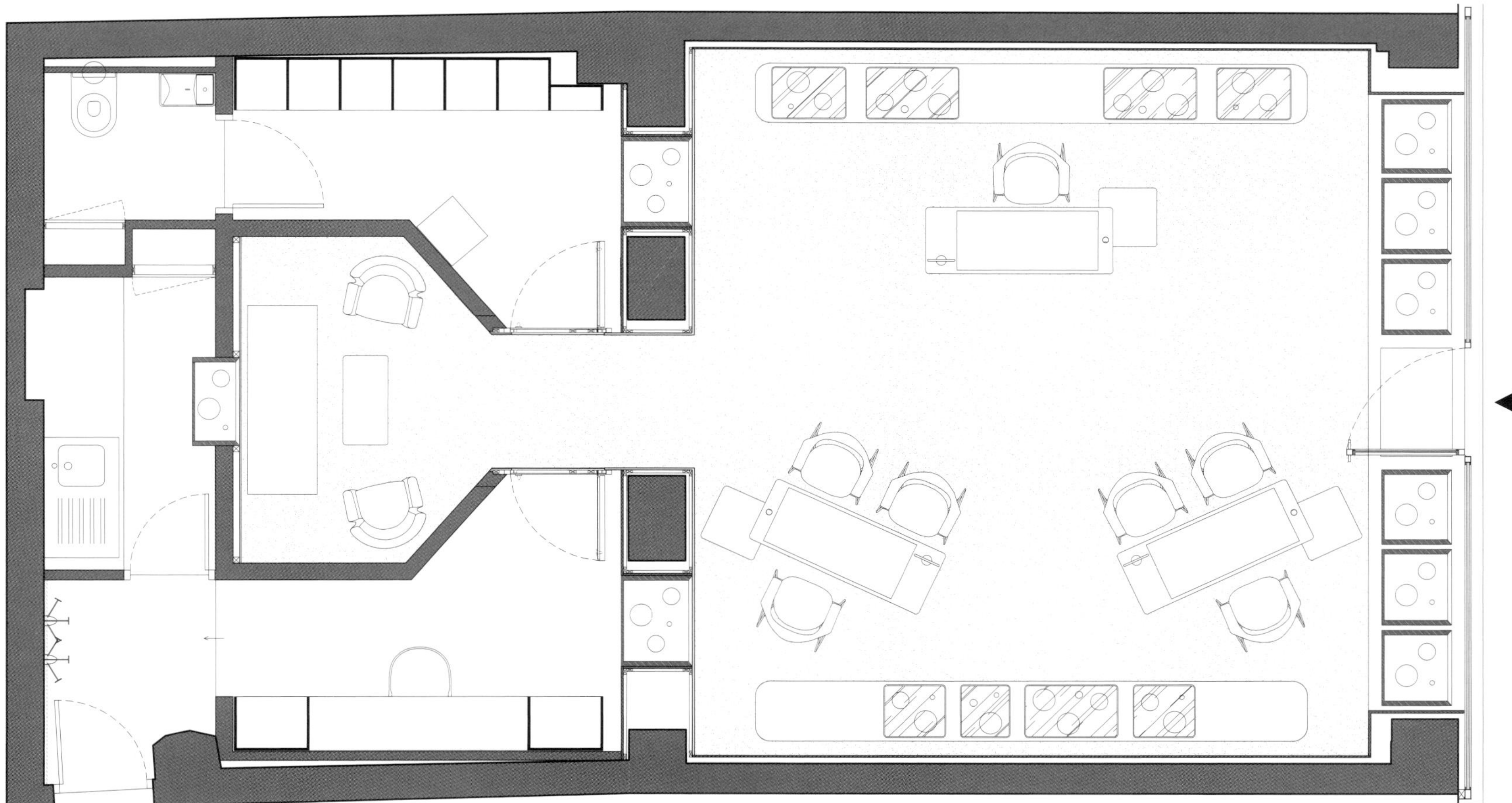

Bulgari Store

Design Agency:
Studio Marco Piva

Like a Bulgari jewel born of a drawing made with water colours or tempera, and then formed into a perfect object by expert craftsmen, hand drawn sketches represent the initial image and expression of the concept of beauty that lays under Marco Piva's projects, which stroke after stroke take shape to become a concrete and vital element.

The inspiration stems from the desire to make a great theatrical exhibition, different and innovative, able to showcase the aesthetics and manufacturing techniques of Bulgari creations in a fascinating setting of shapes, lights and colours, rich with references to the history of goldsmithing and Italian design.

Intensive attention to detail and the rigorous selection of exclusive materials have contributed to the creation of a new window dressing concept for Bulgari that emphasizes the quality and original, unmistakable style of this brand, one among the most innovative and significant in the global jewellery industry.

In Marco Piva's project for Bulgari, one feels that the soft, realistic shapes of the exhibition elements have a clear artistic reference to the portraits by Modigliani, with the sublimely beautiful women's necks and faces, suspended in an eternal moment of elegance and expressed through the immaculate brilliance of ceramics (a material that has been used in the past by Bulgari, with great success) and the sheen of metal that defines the thin shelf supports and light sources as if it were molten gold.

The jewel in all its forms, materials and colours is thus glorified and is displayed so that it can be admired and desired in all its perfection.

The materials selected for the window dressing are metals, woods and ceramics embellished through the use of refined techniques, the result of a millennium of unequalled craftsmanship made current with the latest technologies.

The tubular elements supporting the lamps and the shelves are made from a special aluminium alloy whose surface has been treated with a finish of sparkling gold, the gleaming colour of luxury and class.

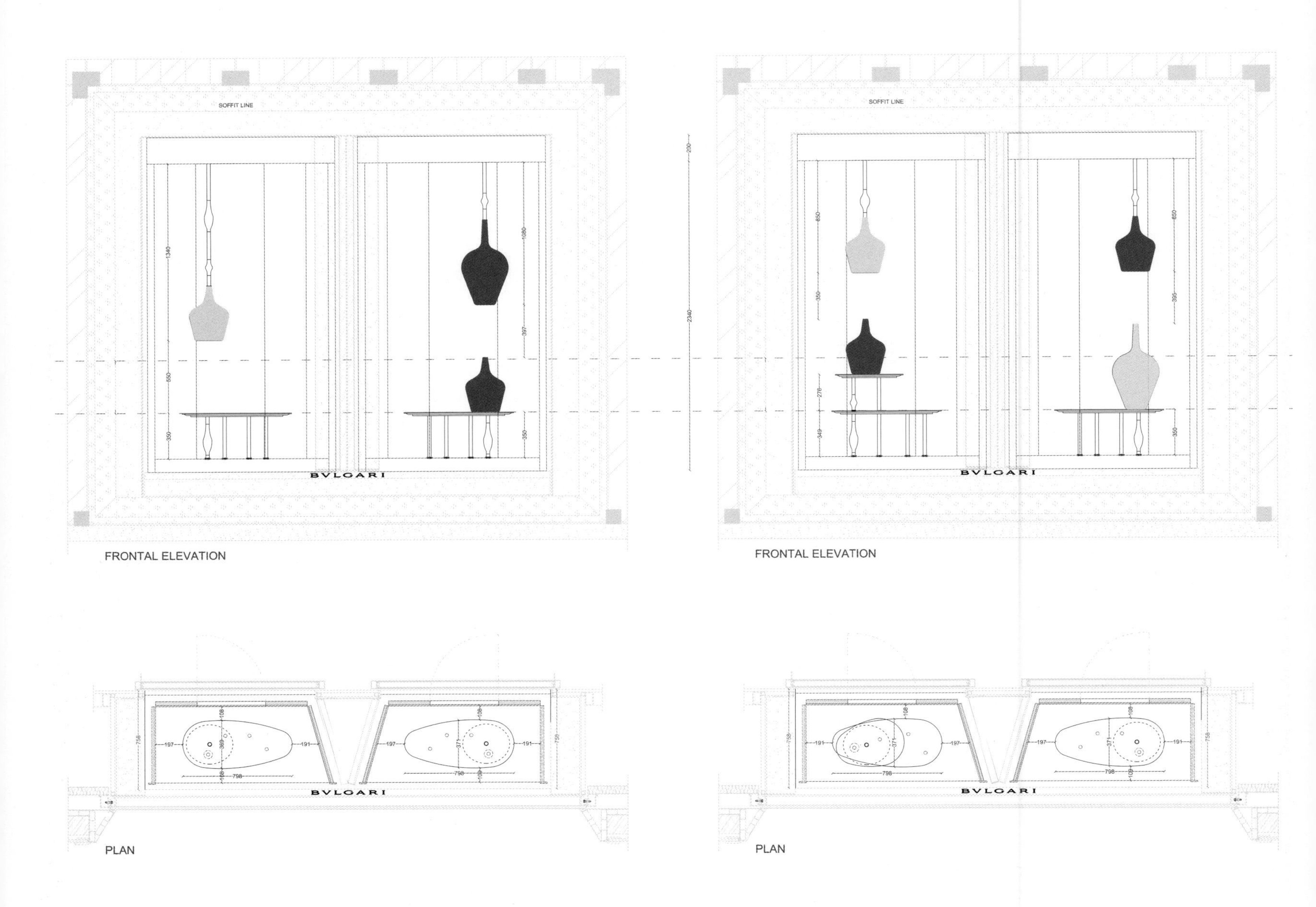

Lacquered wood has been used to create the shelving to reflect and diffuse the light in the exhibition space.

Ceramics were used for the lamps and props, this being a material that symbolizes the strength and quality of Italian craftsmanship, as it can be shaped and used in such a way that the end result can vary from a prestigious work of architecture to a tiny, precious piece of jewellery.

In the window dressing created for the Biennale des antiquaires at the Grand Palais in Paris, Marco Piva's design is pushing the ceramics to the limits of its manufacturing potential by creating what are perhaps the largest shapes ever moulded, simulating a small crowd of elegant figures in the round, reminiscent of Fausto Melotti sculptures, with the unusual life size elements showcasing the most exclusive jewels of the Bulgari collection.

Light is also of the essence in this new window dressing: the large ceramic lamps with their organic and encompassing shapes are both exhibition elements and light generators, while the theatrical use of fabric in the windows interacts with the light to create the unique and exclusive atmosphere of the Bulgari world.

Shanghai KELA.CN Flagstore

Designer:
Zhang Tao

Location:
Shanghai, China

Client:
KELA.CN

Shanghai KELA.CN flagstore is a premium exclusive store in Nanjing East Street, one of the busiest commercial streets in Shanghai.

The interior space is divided into 3 function parts: the big sales area, the sales service area beside the window, and the backup area which is relatively separated. The corporate color of KELA.CN is applied as the stores general color. For the materials, the ceiling is designed as the combination of aluminum strips and modeled plasterboard which meets the requirement of the shopping mall; the floor chooses the collage carpet, relying on its different color pattern to differentiate the custom area and service area. The wallpaper and wood strip decorated walls show an affinity for the customers.

Professional projection light and LED linear light are installed together in the main zone and the spotlighting is used in the secondary zone. The different usage of the light generates a diversified effect for this space.

The whole design is clean and approachable with elegance.

KELA.CN
KELA.CN

KELA.CN
KELA.CN

KELA.CN
KELA.CN
KELA.CN

ELECTRIC STORES

AER – Flagship Store

Design Agency:
COORDINATION ASIA

Location:
Shenzhen, China

Photography:
COORDINATION ASIA

Client:
AISIDI

Area:
130 m^2

DRESS
IT
UP
APP
UP
YOUR

TAP
APP
UP
YOUR
Life

COORDINATION ASIA completed a new breed of telecom stores named AER for AISIDI, one of China's leading resellers for mobile and digital products and services. AER is a retail brand that enhances the life of the individual mobile user by offering customized mobile services in a playful, cool and customer-focused environment. COORDINATION ASIA took on branding as well as VI and store design for AER. The first AER store has now opened in Shenzhen and new stores are planned to open China-wide within the coming year.

Inspired by the growing importance of mobile Internet use in daily life in China, COORDINATION ASIA created an out-of-the-box brand that turns the act of purchasing a mobile product into an active and fun experience. AER is based on a great understanding of mobile lifestyle, in which mobile devices keep you connected, entertained and updated through a variety of online and offline apps.

The store is designed as an interactive environment that caters to the needs of different target-customers: Trendy, Lifestyle and Tech Savvy. Products are thematically presented in combination with related accessories, apps and carriers on custom-made presentation tables with "serving trays". Following a black runway from the entrance, customers find the App Bar where they can try out mobile apps on a large interactive screen. Painted pegboards are used to cover the walls of the store. This allows display shelves to be hung freely within the whole space and to be changed location at all times. The whole interior is flexible, so that changes in layout can be made quick and easily.

COORDINATION ASIA has created a concept that fits perfectly the needs of the mobile-minded customer, who is used to "click and play" and never wants to be bored. The ever-changing store concept of AER will keep customers coming back, providing them with an exciting and new experience every time.

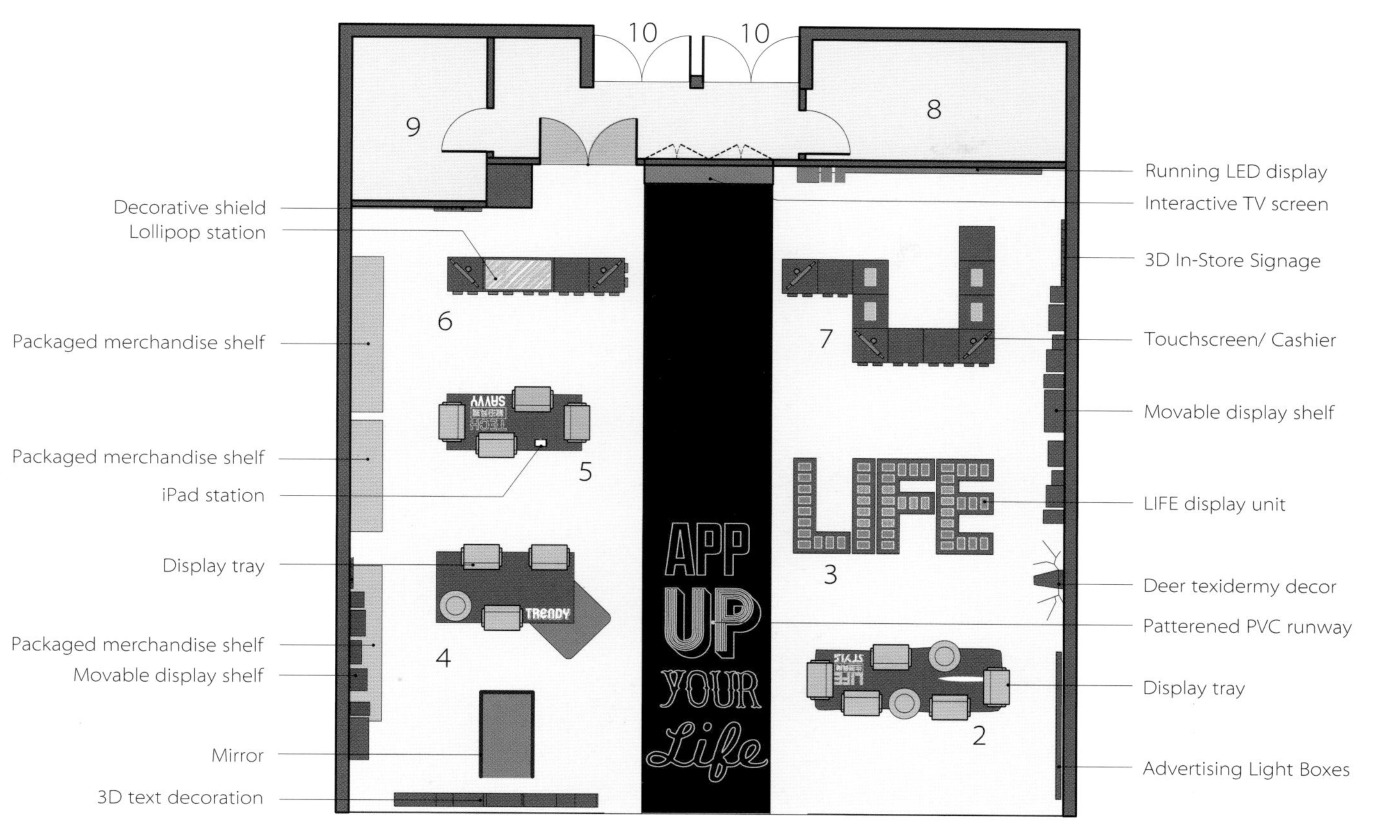

1 Entrance
2 LIFESTYLE table
3 LIFE shaped display unit
4 TRENDY table
5 TECH SAVVY table
6 Service counter
7 Service counter
8 Storage
9 Office/Storage
10 Emergency Exit
Total:130 m^2

Groupon Concept Store

Design Agency:
HEAD Architecture and Design Limited

Designer:
Mark Panckhurst

Client:
Groupon

Location:
Causeway Bay, Hong Kong, China

Area:
1,700 m^2

Groupon (a portmanteau derived from "group coupon") is a deal-of-the-day website that features discounted gift certificates usable at local or national companies.

It started as an online shop and customers could purchase products and service online with discounts and afterward redeem with Groupon's coupon at the merchants' stores. For the convenience of all customers, Groupon has opened their 1st Concept Store in Causeway Bay, Hong Kong to allow a 1-stop for redeeming all products they purchased and no longer need to visit several stores for multiple purchases. The Concept Store also provide a new shopping experience which allow online purchasing with the Terminals – iPads & iMacs and redeem in the store at once.

HEAD Architecture has designed a welcoming space for Groupon Concept Store with glass wall and long benches to rest on while they are queuing for their product. Groupon logo has been used as the featured ceiling lights in the shop. The brand is enhanced further by using the brand color in the whole shop.

on Concept Store !

EXIT 出口
Pioneer

Nokia Flagship Store Helsinki

Design Agency:
Sundae Creative

Location:
Helsinki, Finland

Client:
Nokia

Area:
200 m^2

The client brief was to produce a retail space, which pushed the envelope of normal technology stores both through its visual language and the range of activities and customer interactions possible within the store.

The monochromatic pallet was part of the client requirements, designed to showcase the color ranges available with in each Nokia product family.

Sundae Creative's response was to produce a white landscape of lit wall bays and ribbon like tables: The wall bays are made up of illuminated pill shaped drawers, each capable of displaying devices and accessories and each adaptable to create different display solutions. Every drawer is lit around its edge and when open its face becomes illuminated too to highlight the product. The effect is a futuristic vertical landscape.

The linear, ribbon like tables showcase product demos, each explaining the key features of devices within the Nokia range using experiences based around imaging, navigation and music to encourage customer interaction with the products.

Retail ID for the product families is displayed via vertical LED panels, each with custom designed video content. A central campaign wall uses a mixture of moving and static campaign graphics.

LUMIA

CHARGE IT
THE WORLD'S MOST INNOVATIVE SMARTPHONE.
THIS IS LUMIA.
Claire Young
SkyDrive

THE
WORLD'S MOST
INNOVATIVE
SMARTPHONE.
THIS IS LUMIA.
WITH
NOKIA
CITY
LENS
ASHA
LUMIA
LUMIA 920
Windows Phone

THE
WORLD'S MOST
INNOVATIVE
SMARTPHONE.
THIS IS LUMIA.
WITH
NOKI
CITY
LENS
Claire Young
TEN

OTHERS

Kubrick Bookshop and Cafe

Design Agency:
One Plus Partnership Limited

Design Team:
Ajax Law Ling Kit, Virginia Lung

Location:
Beijing, China

Area:
320 m^2

Photography:
Ajax Law Ling Kit, Virginia Lung

Inheriting the simple but connected idea of the adjacent MOMA bookshop, the café space has enjoyed the eye-popping design concept that the contemporary vibrant blended with modern thoughts. The grey, stable self-leveling screed space is filled with shocking green joinery that works as bookshelves, cashier counters and stock storage, they even developed into the randomly arrayed display cabinets that suspended from the ceiling. The long curvy strip of these cabinets is a materialized response to the hanging walkways outside, as well as the spirit of connection represented by them. The choice of the theme color is mainly concern of the "Five Elements" notion ancient Chinese tradition, the robust green hue in one of those represents the wood element, which is the material of books on the other hand. Despite of their bright green finish, the joinery are formed after stacks of books, which implied the meaning of the accumulation of knowledge through generations. The knowledge contains Beijingers' past, present, and the way to create future.

The extending network of the green electric cords, and the bookshop's green and black lightings suspended beneath it, remind visitors of the messily spread power lines in old alleys all over China. These lines connect to every family in the country, and they live upon these lines; just like books, they connect not only families, but also generations, through the heritage of knowledge. Another approach of the design of the lighting is its linkage with movies, which leads to the scientific usage of optical lights, and the chronicles of their inventions.

According to the above, most part of the design advantages mentioned are based on warmth of natural that can be easily found in timber substance, such as the wooden feature wall and all the furniture pieces containing timber. These also provoked the contrast of soft and hard when they accompany the cool concrete floor.

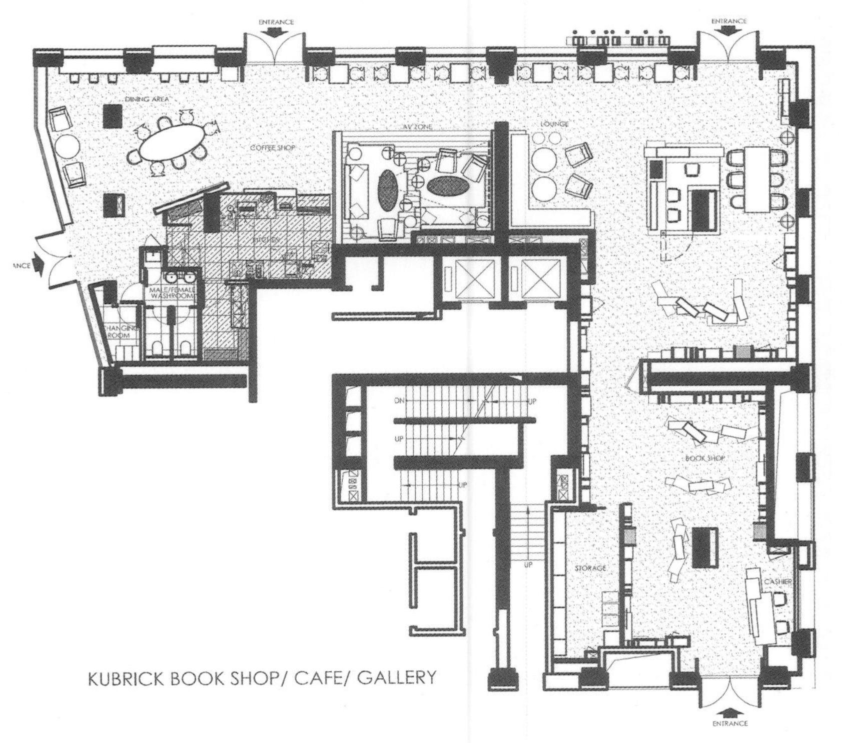

KUBRICK BOOK SHOP/ CAFE/ GALLERY

Fender Custom Shop Mexico City

Design Agency:
Arquitectura en Movimiento Workshop

Location:
Mexico City, Mexico

Area:
100 m^2 covered area
100 m^2 exterior area

Photography:
Cuauhtémoc García

Fender

MARQUEE MOON PSYCHO
COMFORTABLY NUMB
THEME FROM THE DUKES OF H
RADIO RADIO ANAR
HANGING ON THE TELEPHONE
SOMETHING TO TALK ABOUT LON
MESSAGE IN A BOTTLE
BRASS IN POCKET MY BE
(WHAT'S SO FUNNY 'BOUT) PEACE,
WALKING ON THE SUN
BURNING DOWN THE HOUSE
WAITING ON THE WORLD TO CHANGE
WELCOME TO PARADISE
EVIDENCE SHINE ON
& BOYS ROXANNE
ME SMELLS LIKE TEEN
THE STREETS HAVE NO NAME
I STILL HAVEN'T FOUND WHAT I'M LOOKING
GLORY DAYS WISH
SEVEN NATION ARMY
TO FORGIVE BITTER SWEET SYMPHONY
YOU CRAZY DIAMOND
CRUEL TO BE KIND CALL ME
CUTS LIKE A KNIFE LITHIUM
SPIRIT RACE FOR THE PRIZE
KARMA POLICE SUMMER
FOR ALL THE SMALL THINGS

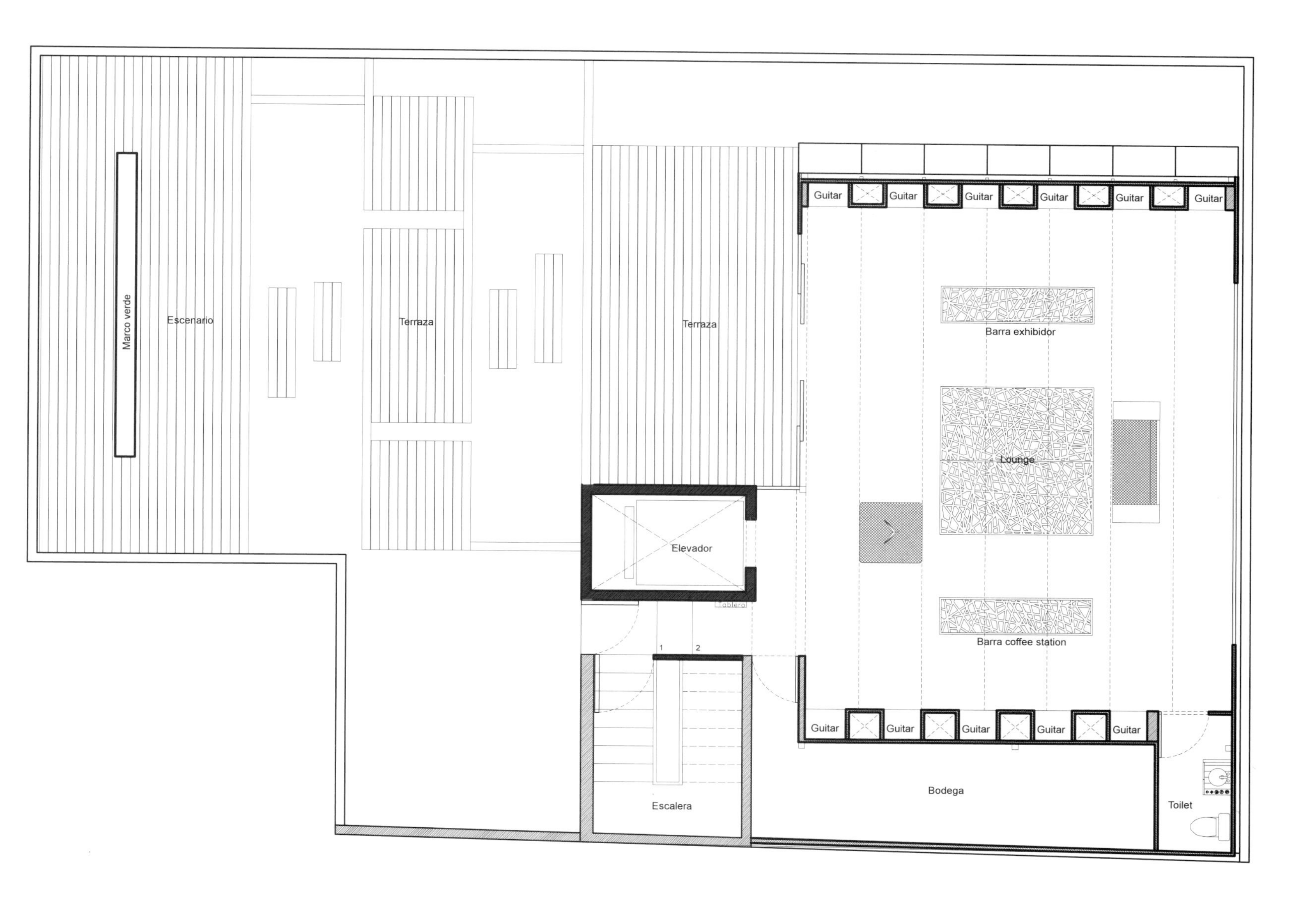

planta arquitectonica

Located within the historic center of Mexico City, on the rooftop of one of the most important themed stores of the music scene, the Fender Custom Shop Mexico City is Latin America's first Fender store and follows a different concept from those in other countries.

The basic concept for the development of the project was an architecture conceived as temporary, a sort of hybrid space that could function equally as a store, an art gallery, a meeting space or a lounge; together with music, a short-lived happening and the idea of the guitar as an anthropometric unit of measure.

The volume in the form of an irregular polygon represents the movement of the vibration of music. The roof is composed of eight sections which slope in different directions and at different angles, generating patterns of natural light in the interiors. Lighting, a key element in the design process, was conceived according to use and function. The mass of the building appears on the rooftop as if sitting on light.

Inside, the project comprises a guitar display area, a lounge, a storeroom and restrooms. The exterior is a timber and stone platform with a level stage, framed by means of an arc of vegetation that directs the eye towards the city vistas. The immediate urban context, however, lies behind a balustrade of asymmetric motifs that serves as a visual barrier. The elevator shaft is the vertical axis around which the different areas are distributed.

The reduction of the original budget by two thirds did not affect the initial design. Instead, it prompted experimentation with simplification of the materials used and construction methods employed. The timeframe for completion was three months.

The building is a metal structure clad in Multypanel. The sturdiness of this material, its acoustic and thermal properties and its low maintenance make it a suitable choice of finish. The simplicity of the interior floors and walls in white-painted pinewood, together with the exposed structure in matching colors – integrated rather than concealed – are in sharp contrast to the brutalistic exterior painted dark.

Integral to this project are furniture design and the choice of materials: in the walls and display niches, the armchairs finished in acoustic foam and a counter in high-shine white polish. The feature ceiling panels use an asymmetrical pattern that is a repeat of the external balustrade, the graphics on the store front suggest tattoo art.

COEO – House of Good Deeds

Design Agency:
dan pearlman Markenarchitektur GmbH

Designer:
Marcus Fischer

Location:
Berlin, Germany

Area:
524 m^2

Photography:
diephotodesigner.de

in guter Gemeinschaft
coeo
MITTWOCH
01
DEZEMBER
schaft
gute Lektüre
Shoppen und dabei Gutes tun

"Combining beauty with goodness" was the motto for the opening on 19 November 2010 from the house of good deeds in Berlin. The new retail concept is a "pilot project" because it combines products from four different sectors under one roof: In addition to fair trade products and goods produced lovingly in workshops for people with disabilities, there are a generous selection of books and a café on more than 500 square meters which invite everyone to stay.

COEO is a store concept for sustainable consumption, which establishes the issue of corporate social responsibility to the principle of the business. The modern design and the architectural concept of the store divide the spacious retail space in four areas: In Good Community, Well done, Fair is good, and Good reading. Each of these four areas has its own color.

Besides the entire retail design, dan pearlman developed the brand and its values, the brand name, the logo, the corporate design and the office equipment.

am Gutes tun
coeo

coeo
Haus der guten Taten
gute Lektüre
COEO mag gute Nachrichten. Neben Büchern zu Lebensfragen und Spiritualität lässt ein großes Sortiment an christlicher Literatur und anderen Medien keine Wünsche offen.
gute Lektüre

Tiny Footprints

Design Agency:
Liquid Interiors

Client:
Caroline Williams

Area:
149 m^2

Design Team:
Rowena Gonzales

Location:
Central, Hong Kong

Photography:
Leo Leung, Jeff Floro

Imagine an inviting haven tucked away on a nondescript 10th floor. This lifestyle destination, located on Duddell Street, is Hong Kong's first Eco baby shop designed with new moms in mind. The name Tiny Footprints was specially chosen not only suggests a focus on children, but ecological responsibility as well.

This city is an unfriendly place for mothers, with very few public facilities for baby changing and virtually no space for breast feeding. Liquid Interiors worked with Australian mom and pilot wife, Caroline Williams in creating a unique one stop eco-baby shop. Here, moms are welcome to feed their babies in a comfortable and soothing atmosphere.

Everything about this shop is cozy and inviting. The colors are muted and even the walls are made of soft fabric. Whimsical trees and clouds provide a calming escape from the busy city just outside. Children's merchandise is within their reach and eye-level. Toys are on display in beautiful jars just like a candy shop. Every bit of the shop has been created with mother and child in mind.

Along with aisles of various baby products, Tiny Footprints features a play and reading area. The retail shop also transforms into a learning center. Educational workshops for parents take place regularly. The shop is equipped with a pull-down projector screen, AV system, and flexible furniture arrangement. Some of the furniture is on wheels, and the display tables were designed to also function as benches as needed. It is the ideal space for learning, growing, and socializing. It is also a comfortable space for moms to breast feed with friends. The changing room acts as a relaxing, quiet area for Moms seeking privacy for feeding and diaper changing.

The Eco design promotes healthy living by using low VOC materials, paint, and adhesives. Energy saving elements includes LED lights and air conditioning zoning controls. During construction waste was recycled or reused as well. The shop was created with care and quality in mind, so that we may only leave the tiniest of footprints meant to fly.

TOYS
READ

TINYFOOTPRINTS

HAUS 658

Design Agency:
Malherbe Design

Area:
375 m^2

Location:
Shanghai, China

HAUS 658 is a Hong Kong-based company with a history in artificial flowers, has opened a store in a new luxury mall on the Bund in Shanghai that sells interior products created by famous designers.

The brand collaborated with Malherbe Design to create the premises of the flagship store in China. The whole retail experience is arranged around five key rooms, including the boudoir, the bedroom and the library; each having totally different "colour ambiances", thus creating strong visual surprises for shoppers.

Innovative displays such as floating pillows and bed sheets engage the customer and enhance the overall visual experience.

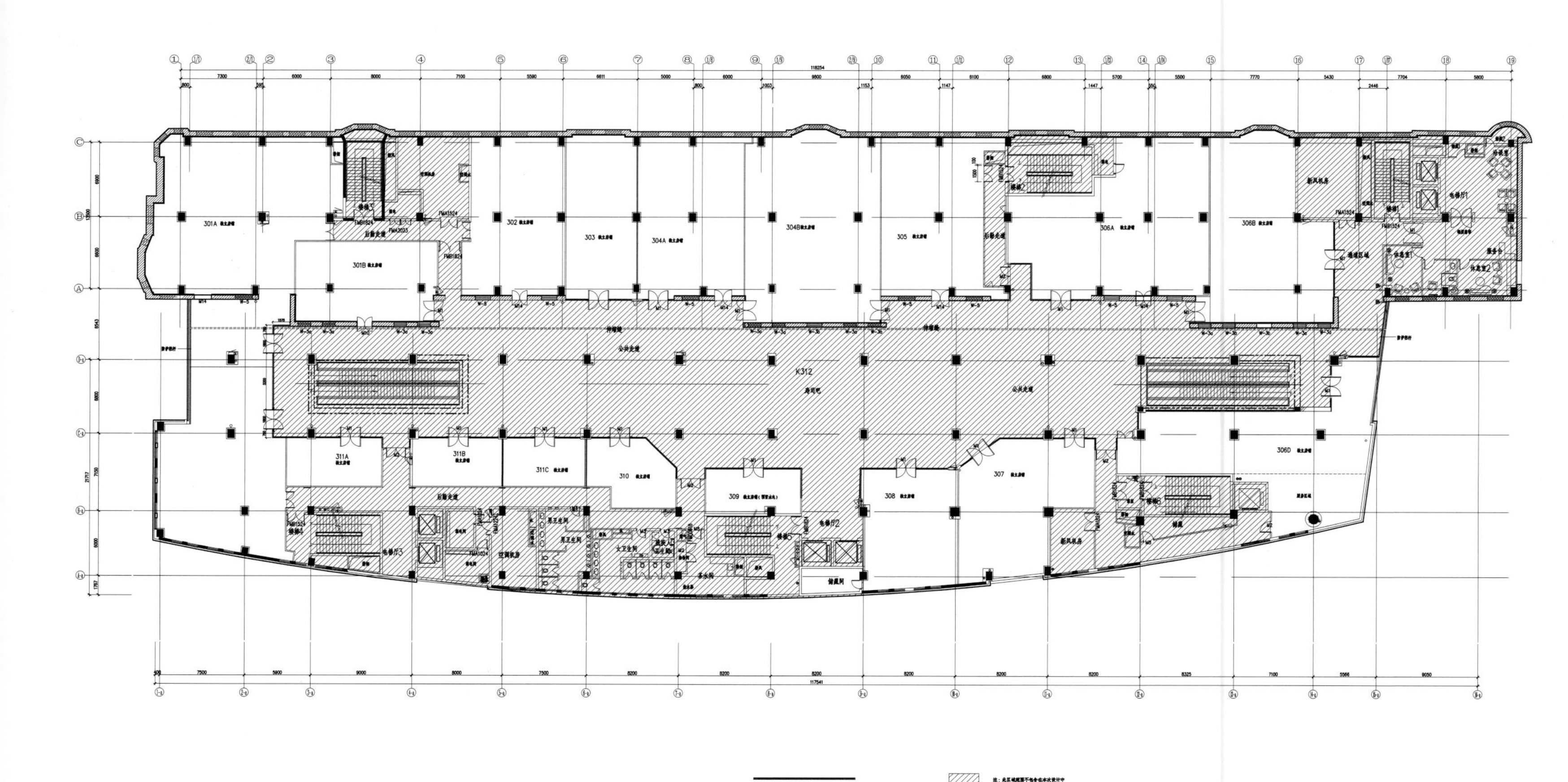

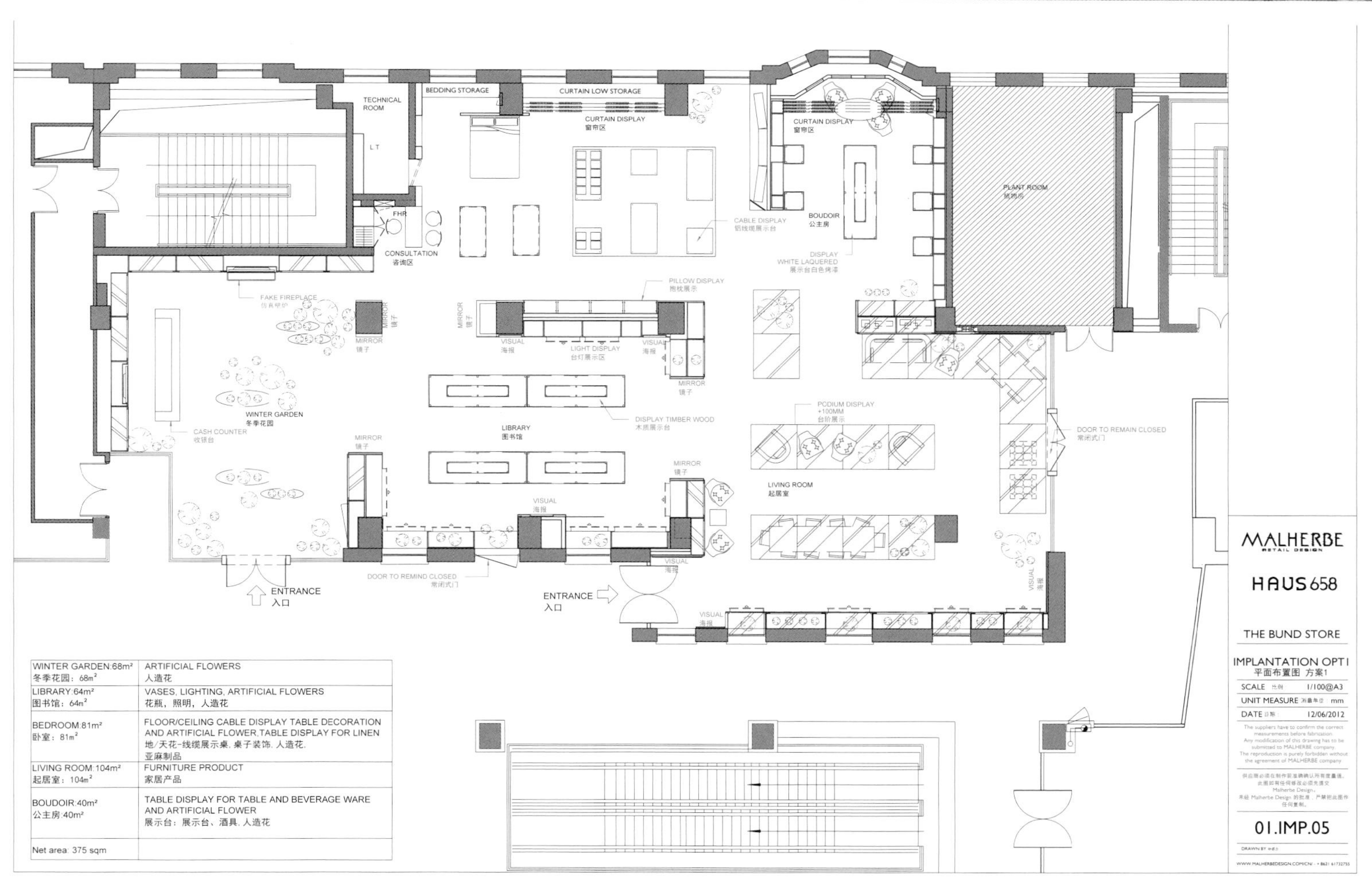

WINTER GARDEN:68m² 冬季花园：68m²	ARTIFICIAL FLOWERS 人造花
LIBRARY:64m² 图书馆：64m²	VASES, LIGHTING, ARTIFICIAL FLOWERS 花瓶，照明，人造花
BEDROOM:81m² 卧室：81m²	FLOOR/CEILING CABLE DISPLAY TABLE DECORATION AND ARTIFICIAL FLOWER,TABLE DISPLAY FOR LINEN 地/天花-线缆展示桌. 桌子装饰. 人造花. 亚麻制品
LIVING ROOM:104m² 起居室：104m²	FURNITURE PRODUCT 家居产品
BOUDOIR:40m² 公主房:40m²	TABLE DISPLAY FOR TABLE AND BEVERAGE WARE AND ARTIFICIAL FLOWER 展示台：展示台、酒具. 人造花
Net area: 375 sqm	

Barbie Shanghai

Design Agency:
Slade Architecture

Design Team:
James Slade,
Hayes Slade

Client:
Barbie

Location:
Shanghai, China

Area:
3,252 m^2

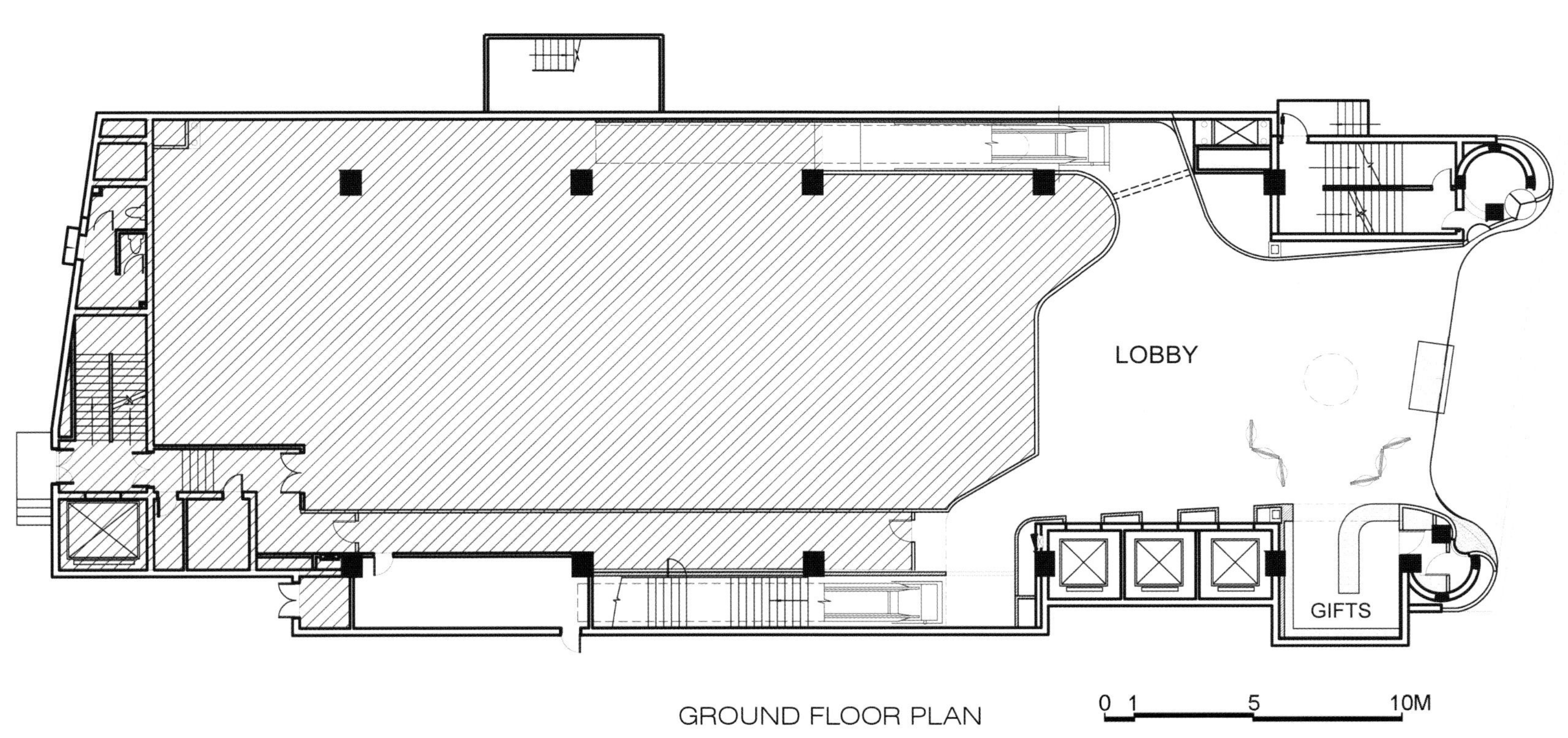

GROUND FLOOR PLAN

The 3,252 m^2 store is the first ever Barbie Flagship. Mattel wanted a store where "Barbie is hero", expressing Barbie as a global lifestyle brand by building on the brand's historical link to fashion. Barbie Shanghai is the first fully realized expression of this broader vision. Slade Architecture led the design of this project, including the exterior, interior, fixtures, and furnishings. Slade's design is a sleek, fun, unapologetically feminine interpretation of Barbie: past, present, and future.

The new facade combines references from product packaging, decorative arts, fashion and architectural iconography to create a modern identity for the store, expressing Barbie's cutting-edge fashion sense and history. The façade is made of two layers: molded, translucent polycarbonate interior panels and flat exterior glass panels printed with a whimsical lattice frit pattern. The two layers reinforce each other visually and interact dynamically through reflection, shadow and distortion.

Visitors are enveloped by the curvaceous, pearlescent surfaces of the lobby, leading to a pink escalator tube that takes them from the bustle of the street, to the double-height main floor. A three-storey spiral staircase enclosed by eight hundred Barbie dolls is the store's core; everything literally revolves around Barbie.

The staircase links the three retail floors: The women's floor (women's fashion, couture, cosmetics and accessories), the doll floor (dolls, designer doll gallery, doll accessories, books), and the girl's floor (girls fashion, shoes and accessories). The Barbie Café, also designed by Slade Architecture, is on the top floor.

Throughout the retail areas, Slade played with the scale differences between dolls, girls and women. They reinforced the feeling of youth and the possibilities of an unapologetically girlish outlook (regardless of age) by mixing reality and fantasy and keeping play and fun at the forefront — to create a space where optimism and possibility reign supreme as expressions of core Barbie attributes.

+NC Store

Design Agency:
COORDINATION ASIA

Client:
Shanghai Glass Co., Ltd.

Location:
685 West Changjiang Rd, Shanghai, China

Area:
240 m^2

Photography:
COORDINATION ASIA

As part of the ongoing collaboration with the Shanghai Museum of Glass, COORDINATION ASIA designed a unique new shop for the museum. In the museum shop the visitor feels like a collector in a colorful glass gallery. After a visit to the museum exhibitions, visitors have a chance to stop by the museum shop to select a special piece to take home with them as a memento. The shop offers a wide range of products that are carefully chosen by the museum team.

The shop design plays with mirrors and both colored and clear glass. Through lights, the glass shines and reflects colors all around and casts dancing refractions in the space. Apart from purchasing precious glass souvenirs, the visitor can sit down and have a coffee at the self-service mini-café, or enjoy a drink to-go while admiring the entrance hall of the museum.

Contributors

Alessia Silvestrelli

Born in 1976 in Ancona, Italy, she received a degree in Architecture in 2004 at Università degli Studi in Camerino.

Her interest is in art, design, and beauty in each of their manifestations orients her expressive research toward interior design. In 2007, she began collaboration with a pharmacy service company, Ataena srl in Ancona, creating commercial spaces in which attention to detail, matching materials, and graphics blend with scientific functional and technological distribution.

ALBUS Design

Henrique Steyer began his academic life studying advertising, and graduated as an architect in 2006. He is post-graduated in Advertising as an Specialist in Advertising Imagery and post-graduated in Strategic Design from POLI.design — Consorzio del Politecnico di Milano. In 2009 he attended the Trends Analysis Course in Milan, Italy. In 2010, he received the Young Talent Hunter Douglas Award, in Turkey. His work has been published in more than 20 countries, including a cover story in Top Decoration World magazine, from China, where his work has been divulged alongside names such as the Egyptian Karim Rashid and the Italian Fabio Novembre.

Arquitectura en Movimiento Workshop

Arquitectura en Movimiento Workshop is formed by young architects, graduates of different universities in Mexico, EEUU and Spain, founded in 1996. Their professional practice has led them to realization of projects of public and private character in different Geographic latitudes and parallels.

Openness and diversity directly influences the team in their creative process, being a reflection of different academic philosophies, which exercises professionally each of its members.

The topics of these projects, allow the exploration and approach to new technologies, reactivating strategies and environmental inertias, which allows a better understanding and utilization of the immediate environment.

In 2012 they received an honorary mention given by the “XI Bienal de Arquitectura Mexicana” in category of offices for their project “Corporativo GA”, located at Juarez, Chihuahua. Also there have obtained a 2nd place in the “Premio Nacional de interiorismo”.

Atelier du Pont

Founded in 1997 by Anne-Cécile Comar, Philippe Croisier and Stéphane Pertusier, Atelier du Pont is a multipolar agency that oscillates between public and private projects, architecture and urban planning, exterior and interior work.

Although architecture – and in particular producing public housing and facilities – is an essential part of the agency's work, interiors are a breath of fresh air that gives free rein to their imagination, allows to experiment with extravagant materials and to create places that are both outside time and of their time.

Beijing Matsubara and Architects

Beijing Matsubara and Architects (BMA) was established in 2005. The office is currently based in China, practicing urban, interior and architectural design.

BMA is highly regarded architecture office producing inventive and sincere design, and also high quality construction management, especially appreciated from the past projects, such as Liuliqiao Office, Xinhua Bookstore Zhongguancun Book Building, Courtyard by the Canal, Sanlitun VILLAGE North Area EAST, Royal tea Barn, and Y house. They keep exploring the possibility of architecture, and produce more superior projects that can be appreciated for future, at the same time aspiring to become a bridge between Japan and China.

bluarch architecture + interiors + lighting

Antonio Di Oronzo is a Professor of Architecture at City College of New York, School of Architecture, Urban Design and Landscape Architecture. He also teaches interior design at Parsons School of Design, School of Constructed Environments. He has taught graduate courses in digital culture and aesthetic, and media design at New School University, Department of Media Studies and Film, and branding and brand management at Parsons School of Design, Department of Design and Management.

In 2004, Antonio founded the award-winning firm bluarch architecture + interiors + lighting, a practice dedicated to design innovation and technical excellence providing complete services in master planning, architecture, interior design and lighting design. At bluarch, architecture is design of the space that shelters passion and creativity. It is a formal and logical endeavor that addresses layered human needs. It is a narrative of complex systems which offer beauty and efficiency through tension and decoration. Based in New York City, the firm is recognized for both built and speculative work in both publications and exhibitions.

Concrete Architecture Associates

Concrete Architectural Associates was founded in 1997. Rob Wagemans is founder and creative director of Concrete Architectural Associates, he was born in eindhoven on 13th of february 1973. He got his bachelor degree in architecture utrecht, master degree of architecture at academy amsterdam.

Concrete consists of 5 fundamental building blocks: concrete interiors, concrete architecture, concrete tomorrow, concrete today and concrete heritage. Concrete's entire team consists of about 35 professional people. Visual marketers and interior designers, graphic designers and architects work on the projects in multidisciplinary teams.

The house of concrete is based in a listed building in the middle of the red light district on the oudezijds achterburgwal. Here, the designers work, in an interdisciplinary way, on the total concepts in brainstorm sessions. The company builds brands, produces the interior design, architectonic and urban development plans, along with the main presentations and, eventually, the scale models themselves.

Concrete develops total concepts for businesses and institutions. The agency produces work which is commercially applied. This involves creating total identities for a company, a building or an area. The work extends from interior design to urban development integration and from the building to its accessories. Concrete, for example, also sets the perimeters for the graphic work and considers how the client can present itself in the market.

This all happens from the “one concept” philosophy. The designers of concrete create holistic plans and everything they design is used for the benefit of that total concept: that's where their strength and thus the client's greatest advantage lies.

Concrete is dynamic, quick on its feet and self-determined. The agency thrives on hard work and the creation of beautiful things. Concrete does not have a pre-determined style and the designers do not simply create designs, interiors or buildings: concrete devises solutions.

COORDINATION ASIA

COORDINATION ASIA is an award winning agency for design and architecture, specializing in creating high impact environments in Museum, Exhibition, Retail and Hospitality Design. Driven by the quest for fresh ideas and extraordinary design solutions, COORDINATION ASIA functions like a concept laboratory and bridges creative energy and professional realization.

From cultural to commercial projects, COORDINATION ASIA's vision is that content precedes aesthetics and that great design is more than simply styling. The office is known and appreciated for this content-driven approach, as well as for a dedicated, reliable and high-standard way of working. This has resulted in successful collaborations with amongst others the Shanghai Museum of Glass and the Shanghai Film Museum.

COORDINATION ASIA is founded and managed by German architect Tilman Thürmer, who also co-founded COORDINATION Berlin together with architect Jochen Gringmuth and product designer Flip Sellin. The client base of the COORDINATION studios involves global brands such as Deutsche Bank, Nike, Adidas, Falke and Otto Bock HealthCare.

dan pearlman

Marcus Fischer is Creative Director Brand Experience & Exhibition | Founder & Associate of dan Pearlman.

Marcus Fischer was born in 1971. International and national trade fairs stands and exhibition concepts are evidence of his lead management. After receiving his degree in interior design in 1999 at the HDK in Berlin, he founded with three partners dan pearlman, a strategic creative agency, working interdisciplinary across the areas of brand and leisure, strategy and implementation. Marcus Fischer is responsible for clients such as smart, BMW, MINI, Atkon on behalf of Deutsche Bahn International, Roca, Museum of Arts and Crafts in Hamburg, Lufthansa or MTV.

Dalziel+Pow

Dalziel+Pow is an integrated design consultancy, offering a full range of design services, from brand positioning, identity design and retail design, through to graphic design, photographic art direction and website design. Developing brand environments and communications across all key touch-points, the services offered are a holistic approach, which results in successful brand environments which has been designing for its clients since 1983.

The key to success is understanding how customers interact with brands. The team design with the customer in mind, driving brand awareness, product interaction and ultimately, sales.

The offices are located in London, Mumbai and Shanghai. Since 2008 when the Shanghai office was set up, Dalziel+Pow has been talking with retailers based in China for many years. The dedicated Shanghai office ensures international design concepts are correctly translated and implemented to suit the China market, while providing access to the best international creative thinking.

Dalziel+Pow creates retail brands and branded environments with strategic goals that deliver real commercial benefit to clients.

Fourfoursixsix

Fourfoursixsix is an international architecture practice with offices in London and Bangkok. With national and international consultant and procurement relations, they are ideally positioned to provide realisable design solutions in the UK and abroad. Their design process and ability to work within highly varied typologies, coupled with their international positioning, has allowed the company to undertake a diverse range of worldwide commissions within a number of project disciplines.

They continue to carry out research based work and actively pursue competitions in order to further develop our ideas, approaches and philosophies.

Fourfoursixsix are driven by a desire to create innovative, unique and pragmatic architecture. Avoiding the constraints of a "house style", they choose instead to formulate design solutions through an investigative studio process, underpinned by a high level of client consultation and careful investigation of context, site and brief.

Such a methodology allows the firm to work at a wide range of scales and typologies, continually evaluated through rigorous critical analysis. They believe that this process-led approach to design produces consistently high quality solutions for both the client and end user.

GRAFT Gesellschaft von Architekten mbH

GRAFT was established in 1998 in Los Angeles, California by Lars Krückeberg, Wolfram Putz and Thomas Willemeit. Further offices followed in Berlin, Germany in 2001 and Beijing, China in 2003, with Gregor Hoheisel as Partner for the asian market.

GRAFT is a "Label" for Architecture, Urban Planning, Design, Music and the "pursuit of happiness". Since the firm was established, it has been commissioned to design and manage a wide range of projects in multiple disciplines and locations. With the core of the firm's enterprises gravitating around the field of architecture and the built environment, GRAFT has always maintained an interest in crossing the boundaries between disciplines and "grafting" the creative potentials and methodologies of different realities. This is reflected in the firm's expansion into the fields of exhibition design and product design, art installations, academic projects and "events" as well as in the variety of project locations in Germany, China, UAE, Russia, Georgia, in the U.S. and Mexico, to name a few.

Its collective professional experience encompasses a wide array of building types including Fine Arts, Educational, Institutional, Commercial and Residential facilities. The firm has won numerous awards in Europe as well as in the United States. With a staff of talented architectural professionals and administrators, GRAFT has the resources and technology necessary to execute a project from programming to design and through construction, including construction documents, construction administration, and governmental agency review phases. GRAFT has rigorously undertaken an increasing role in programming, master-planning and urban design.

Hangar Design Group

Hangar Design Group is an international creative firm founded in 1980 in Italy by the architects Alberto Bovo and Sandro Manente.

Right from the start, the aim of the studio was to gather under one name the various departments dedicated to communications, graphics, retail, industrial design and branding strategies, conveying different disciplines into a unique design approach.

Today operating in Europe, Asia and America with four offices (Milan, Venice, Shanghai and New York), Hangar Design Group is a multidisciplinary company with its headquarters in Italy, boasting more than forty people and serving international companies worldwide.

HEAD Architecture and Design Limited

HEAD Architecture and Design Limited was established in Hong Kong by a group of Architects, designers and project managers who shared the common goal of the pursuit of excellence in architectural design. Their scope of experience broadly covers all aspects of projects from inception through brief development, conceptual and developed design on a wide range of projects.

HEAD Architecture and Design Limited is headquartered in Hong Kong and managed by Directors Mark Panckhurst and Mike Atkin. With affiliated offices in Shanghai, Brisbane, London and Guangzhou, HEAD are well positioned to respond to assignments around the world.

INBLUM Architects

INBLUM Architects (Vilnius, Lithuania), a multiple winner of interior design competitions both at home and abroad, is a tandem of architects Dmitry Kudin and Laura Malcaitė.

The architects create atmospheric architecture, sensitive to environment, conditions and ways of modern people. These design principles give a solid foundation to the aesthetics they generate, where their architectural message becomes a text open for interpretation. Spiritually close to Modernism, the architects' works focus on a project's content, searching for unconventional, perfectly functional solutions. The architects do not seek after trendy conceptual approaches or unjustified architectural forms. In this material world, they try to reveal the poetic nature of architecture.

Ippolito Fleitz Group

Ippolito Fleitz Group is a multidisciplinary, internationally operating design studio based in Stuttgart. Currently, Ippolito Fleitz Group presents itself as a creative unit of 37 designers, covering a wide field of design, from strategy to architecture, interiors, products, graphics and landscape architecture, each contributing specific skills to the alternating, project-oriented team formations. Their projects have won over 160 renowned international and national awards.

José Carlos Cruz - Arquitecto

After several years in partnership with other architects, Jose Carlos Cruz created his own company in 2004. Frequent visits to major world cities offer a wide platform of knowledge focused on ways of approaching fundamental problems of space, volume, proportion, light, material, details and furnishings. The office remains true to the practical realities and lifestyles of each client. It seeks to give each project an individually character tailored to each site, making in unique and personal, as well as to look for realist solutions together with the client. The office's works continues to be published in international magazines and books such as Frame, Attitude, Area, Hauser, Arquitectura & Construção, Chain Reaction, Architecture y Desino, Abitare, Wallpaper, 1000 x European Architecture, 24 houses and Casas Contemporâneas.

KLAB architects

KLAB architecture (kinetic lab of architecture), formerly known as KLMF architects, was founded in 2001 by Konstantinos Labrinopoulos. It is an international group of highly qualified and motivated architects who seek opportunities for creating unique and intriguing urban events. Since the language of Architecture travels across borders, this multicultural team invests on its extremely rich diversity of knowledge, from the largest planning scale to the smallest technical details. Such practice is at the forefront of contemporary international architectural design.

The aspiration of KLAB architecture architects has always been: create as artists and to materialize as scientists. Freedom of inspiration, originality in design and strict project implementation are the principles that drive the office's continuous development. Everyday research is an essential ingredient for high-quality innovative and functional design. Every client is unique and poses a new and interesting challenge for the office members, who engage with him in an active dialogue at every stage of the building process. Every project is tailor-made according to the requirements and aspirations of the clients.

KLAB architecture aims for high-quality architecture, one that broadens the mind and expands architecture's potential through unique approaches.

With a considerable number of highly successful projects in Greece and in South East Europe, KLAB architecture is today strongly positioned to confront any architectural challenge with its unique attitude to design.

Lautrefabrique Architects

The offices of Lautrefabrique (Literally The Other Mill) are located in a former silk mill, a listed building in La Galicière a locality of the village of Chatte, Isère, midway between Grenoble and Valence, and 20 minutes from Valence TGV station.

Jean-Pascal Crouzet, DPLG founded Lautrefabrique Architects in May 2001 and leads a team of 5 to 7 employees.

With international experience, gained through working with reputable clients with very high standards, Lautrefabrique offers a full range of architectural services.

Trained in environmental approaches, it offers a pragmatic approach focusing on promoting straightforward, common sense solutions in both public and private markets.

The scale and areas of Lautrefabrique's projects are varied and allow the expression of architecture that the geographical context of the projects and programs requires.

Local projects consist primarily of the renovation of old buildings, both public and private.

Those carried out in the Middle East are either new buildings or interior architecture programs which correspond to a strong brand image, and are projects for which design excels in the expression of functionality, ergonomics, communication and merchandising.

Liquid

Liquid is a green commercial interior design studio based in Hong Kong. They are known for creating better lifestyles by bringing together green interior environments, people and innovative ideas. With increasing importance on having to take action our future, they are committed in improving quality of life while reducing their impact on the environment. Liquid's commercial interior design services for retail, food and beverage and office space enable companies to strengthen their brands and increase productivity and lifestyles through sustainable design strategies.

Liquid is a transparent element in their work that is apparent in the fluid spaces, smooth work flow and solid branding. Commercial space is an extension of a company's image and philosophy where story is told and human connections take place. They work with clients to develop names, identities, sustainable practice and brand platforms to create places where people can experience, buy and believe.

Liquid also strives to maximize energy efficiency and overall cost reductions through the integrative design approach. This approach allows owners, builders, operators, architects, planners, contractors and the design team to interactively develop innovative solutions for a common goal. This holistic design approach is the most efficient way to achieve sustainable results.

Malherbe Design

Malherbe Design is the European leader in retail design with offices in Paris, Casablanca, Hong Kong and Shanghai, with 20 years of experience specializes in retail design and focuses on improving the shopper experience, hence the commercial efficiency. Creativity is leverage in order to increase the sales performance. For Malherbe, the result is what counts. Customers only see the final result, not all the work behind the scene (marketing analysis, focus groups, strategies and the usual numerous power point presentations...). Creating new generations of shops means saying goodbye to the old methods. Hello to field experiment and ground innovation.

Since its launch in 1992, the agency has never stopped growing. During the four last years, it has doubled its size and created a subsidiary in Shanghai and commercial offices in Casablanca (Morocco) and Hong Kong.

Hubert de Malherbe (CEO, Chief Designer & Artistic Director) is involved in all the major projects. For several key clients of the agency he leads an on-going cross-media artistic direction, percolating consistency, style & content to all the brands equities and messages.

MARKETING-JAZZ

MARKETING-JAZZ is the n°1 leading company in Spain specialized in visual marketing.

Founded by Carlos Aires in 2002, MARKETING-JAZZ focuses on creating new store concepts to improve sales.

Its retail projects include creative and integral store design, branding and communication through window displays and training and providing specialized expertise in visual merchandising.

MARKETING-JAZZ, creative retail design.

One Plus Partnership Limited

Established in 2004 by directors Ajax Law and Virginia Lung, One Plus Partnership Limited is an award-winning Hong Kong-based design firm that aims of providing professional services, included: Interior Design, Architectural, Concept Design, Furniture Design, Graphic Design, and Signage Design.

One Plus finds creative inspiration for each project by selecting a specific theme and incorporating novel and unique design elements in keeping with the selected theme.

SAVVY STUDIO

SAVVY STUDIO is a multi-disciplinary studio dedicated to developing brand experiences that generate emotional links between our clients and their audience.

The team is composed by specialists in the areas of marketing, communication, graphic design, industrial design, creative copywriting and architecture. They also collaborate with talented artists and designers from around the world, a process that allows them to offer creative and innovative solutions with a global competitive edge.

SAVVY STUDIO approaches every project with a comprehensive and open creative process, which facilitates the participation of their clients in every step.

Slade Architecture

Slade Architecture was founded in 2002, seeking to focus on architecture and design across different scales and program types. The design approach is unique for each project but framed by a continued exploration of primary architectural concerns.

As architects and designers, the team operates with intrinsic architectural interests: the relationship between the body and space, movement, scale, time, perception, materiality and its intersection with form. These form the basis of our continued architectural exploration.

Layered on this foundation, is an inventive investigation of the specific project context. Slade Architecture's broad definition of the project context considers any conditions affecting a specific project: program, sustainability, budget, operation, culture, site, technology, image/branding, etc.

Sundae Creative

Sundae is a London based creative collaboration combining architects EvansLundin and digital designers Dead Pixels. They together form Sundae Creative, an agency designed to provide both built and digital solutions for retail environments, giving clients a joined up response to their needs when making modern, intelligent store concepts.

Studio Marco Piva

Marco Piva works in the fields of master planning, architecture, interior design and industrial design. Activities vary from large scale structures dedicated to residential and tourist functions, interior design up to the design of particular products for the residential as well as public areas.

Marco Piva develops master plan projects, residential and office complexes, tourist villages, congress centers, meeting halls, exhibition galleries, thematic exhibitions and urban scenes, as well as constant involvement in the most prestigious events and exhibitions worldwide. Assiduous research into the formal and functional aspects of space, updated technology and materials, developed with great attention to the environment, are the foundations of the planning philosophy of the Studio, where the continuity between architecture and interior design plays a strategic role in the success of the Studio's projects.

The Studio's main offices are in Milan, considered the world capital of industrial and fashion design. Studio's branches are located in Dubai (U.A.E.), Sao Paulo (Brazil), Saint Petersburg and Moscow (Russia), Mumbai (India), Beijing (China) and Doha (Qatar).

TORAFU ARCHITECTS Inc.

Founded in 2004 by Koichi Suzuno and Shinya Kamuro, TORAFU ARCHITECTS employs a working approach based on architectural thinking. Works by the duo include a diverse range of products, from architectural design to interior design for shops, exhibition space design, product design, spatial installations and film making. Amongst some of their mains works are "TEMPLATE IN CLASKA", "NIKE 1LOVE", "HOUSE IN KOHOKU", "airvase" and "Gulliver Table". "Light Loom (Canon Milano Salone 2011)" was awarded the Grand Prize of the Elita Design Award. Published in 2011were the "airvase book" and "TORAFU ARCHITECTS 2004 — 2011 Idea + Process" (by BIJUTSU SHUPPAN-SHA CO., LTD.) and in 2012, a picture book titled "TORAFU's Small City Planning" (by Heibonsha Limited).

The launchroom

The launchroom was founded by Huang Zi in 2007 as a freelance initiative to satisfy his creative needs beyond a 9-5 job. It went on to become a full fledge upstart serving independent boutique brands as well as larger FMCG clients with a focus on retail design and brand experience.

The work coming out of The launchroom (TLR) has since been featured widely in several publications as well as online media including Behance, Retaildesignblog, and Lookslikegooddesign.

TLR is essentially one of the best examples of Huang Zi's unique style of a seamless transition between 2D & 3D design as well as his strong sense of project and creative ownership.

UXUS

Established in 2003, UXUS is a leading strategic design consultancy delivering innovative consumer experience solutions. UXUS is widely recognized for its multidisciplinary approach to design, branding and strategic development in retail, hospitality and interior design.

UXUS inspires organizations to analyze, create and develop ideas, find new revenue streams, increase brand equity and ultimately, customer satisfaction.

UXUS' design methodology is "Brand Poetry" and is defined as "Artistic solutions for commercial needs". Brand Poetry is the perfect balance between emotional connection and commercial result.

URBANTAINER Co., Ltd.

URBANTAINER is an architecture and design firm that creates unique atmospheres and spaces for subculture inhabitation within the city. Various social actions and behaviors are continuously taking place within the built urban environment. They seek this social activity and offer it a presence. Their designs are immediately inhabited by events for people to meet, interact, and enjoy. They give subculture a physical presence within the city.

The urban condition in Asia is a complex vertical system of competition, architecturally and socially. Designers at URBANTAINER believe in a horizontal discovery and structuring of the urban environment; a social structure where unique communities can form and thrive. It is their philosophy to understand these communities and enhance them. Design does not only create physical beauty, but has the power to facilitate wellbeing and interaction.

Zhang Tao

Zhang Tao was graduated from Xi'an Academy of Fine Arts in 1999. He has worked in Haigo Shen International Engineering Consultants Inc., Shimaogroup and K. F. Stone Design International Inc. Canada successively as the principal designer.

Zhang is skilled in creating a comfortable and natural human space by using various techniques, taking advantage of the natural environment, he can make the color and lighting effects present different charms cleverly.

In addition, Zhang can masterly combine design elements with nature amongst the space, which provides people with a modern, fashionable, concise and harmonious environment. It is Zhang's professional spirit to maintain the integrity and unity of design that has seen his projects receive the high praise from both his clients and the industry.

ARTPOWER

About ARTPOWER

PLANNING OF PUBLISHING

Independent plan, solicit contribution, printing, sales of books covering architecture, interior, graphic, landscape and property development.

BOOK DISTRIBUTION

Publishing and acting agency for various art design books. We support in-city call order, door to door service, mail and online order etc.

COPYRIGHT COOPERATION

To further expand international cooperation, enrichpublication varieties and meet readers' multi-level needs, we stick to seeking and pioneering spirit all the way and positively seek copyright trade cooperation with excellent publishing organizations both at home and abroad.

PORTFOLIO

We can edit and publish magazine/portfolio for enterprises or design studios according to their needs.

BOOKS OF PROPERTY DEVELOPMENT AND OPERATION

We organize the publication of books about property development, providing models of property project planning and operation management for real estate developer, real estate consulting company, etc.

Introduction OF ACS MAGAZINE

ACS is a professional bimonthly magazine specializing in high-end space design. It is color printing, with 168 pages and the size of 245*325mm. There are six issues which are released in the even months every year. Featured in both Chinese and English, ACS is distributed nationwide and overseas. As the most cutting-edge counseling magazine, ACS provides readers with the latest works of the very best architects and interior designers and leads the new fashion in space design. "Present the best whole-heartedly, with books as media" is always our slogan. ACS will be dedicated to building the bridge between art and design and creating the platform for within-industry communication.

ARTPOWER
Artpower International Publishing Co., Ltd.

Add: G009, Floor 7th , Yimao Centre, Meiyuan Road, Luohu District, Shenzhen, China
Contact: Ms. Wang
Tel: +86 755 8291 3355
Web: www.artpower.com.cn
E-mail: rainly@artpower.com.cn

QR (Quick Response) Code of ACS Official Wechat Account

Acknowledgements

We would like to thank all the designers and companies who made significant contributions to the compilation of this book. Without them, this project would not have been possible. We would also like to thank many others whose names did not appear on the credits, but made specific input and support for the project from beginning to end.

Future Editions

If you would like to contribute to the next edition of Artpower, please email us your details to: artpower@artpower.com.cn